IEE ELECTRICAL MEASUREMENT SERIES 9

Series Editors: A. E. Bailey
Dr. O. C. Jones
Dr. A. C. Lynch

MICROWAVE CIRCUIT THEORY

and foundations of microwave metrology

MICROWAVE
CIRCUIT THEORY

and foundations of
microwave metrology

GLENN F. ENGEN

Peter Peregrinus Ltd. on behalf of the Institution of Electrical Engineers

Published by: Peter Peregrinus Ltd., London, United Kingdom

British Library Cataloguing in Publication Data

A CIP catalogue record for this book
is available from the British Library

ISBN 0 86341 287 4

Printed in England by Short Run Press Ltd., Exeter

Contents

Preface

This book is an outgrowth of the author's 30+year tenure at the National Bureau of Standards (now the National Institute of Science and Technology). His purpose in writing has been primarily twofold. The first objective has been to collect (from his own work and that of his colleagues) and put in better form those ideas which appear to be of continuing rather than transitory interest. The major objective, however, has been to (hopefully) help 'pave the way' and provide a reference book for the next generation of microwave metrologists. With this thought in mind, the author has attempted to assemble, and present in an orderly fashion, those ideas and concepts for which he found repeated use during his career at NBS. In some cases these are in the form of analytical tools which were developed to simplify the solution of recurring problems.[1] In the absence of such a work, the newcomer to the field is confronted with a bewildering array of published papers, each one of which purports to have contributed something to the emerging art.[2]

To be sure, the historical development of the art is strongly mirrored in the publications of record. Moreover, there is a substantial precedent, in a book of this type, to identify and acknowledge the source of the many contributions to the current art. But if carried to its logical conclusion, the book is reduced to little more than an annotated bibliography. Apart from the historian, the typical newcomer is primarily interested in the current state of the art, and not necessarily in how it emerged. For these reasons the

---

[1]For an example, see the Appendix.

[2]In this context, an interesting suggestion has been made to the effect that, *'We must find a way to throw out, for good and for all, a good fraction of the published scientific literature, · · · '* (L. M. BRANSCOMB: 'Is the Literature Worth Reviewing?' *Scientific Research*, May 27, 1968. (Unfortunately, whatever the merit of this idea, the chances of this becoming a reality are probably minimal because of the insurmountable practical problems in selecting that portion which is worth retention.)

author has followed a somewhat different approach. To be more specific, the author has attempted to provide an integrated picture of the existing art, but apart from a few exceptions, no effort has been made to identify who may have contributed what.[3] The references which have been included are primarily limited to those which provide a more complete treatment of the topic under discussion.

As suggested by the title, the book is divided into two major parts, microwave circuit theory, which includes chapters 1-9, and microwave metrology which comprises the remainder. The introductory chapters (primarily Chapter 2) are intended to provide a simplified outline of the theoretical basis for microwave (as contrasted with low frequency) circuit theory. These are based to a large degree upon the prior work of D. M. Kerns, whose example of careful formulation has been a source of unending challenge and inspiration to the author. Following this, the focus shifts to a development of the scattering notation, and where special attention is given to those features which make microwave metrology different from its lower frequency counterpart.

The second part of the book describes the different experimental strategies which have been devised to measure the parameters associated with the microwave model. From one perspective, some of these are now 'obsolete' by virtue of the introduction of the automated network analyzer. On the other hand, for the serious student of metrology, the strategies are of continued interest even if the methods by which they were implemented are not. In the author's experience, an intuitive mental picture of the system operation can be an invaluable asset, although never a substitute for a careful analytical formulation. An example is the tuned reflectometer. Although for most applications the method is obsolete, an understanding of the associated theory can still provide useful insights into the microwave circuit model, and the operation of other measurement techniques.

---

[3]Although in some cases the source of certain contributions is well defined, in others this would prove a difficult if not impossible task.

Although it is not possible to explicitly recognize the many who have contributed, directly or indirectly, to this work, a number of names are deserving of special mention. That of D. M. Kerns has already been noted. In addition, the author would like to acknowledge the guidance from R. W. Beatty during his early tenure at NBS, and his collaboration with C. A. Hoer in the development of the six-port. In particular the widespread interest in the six-port art is probably due, in a large measure, to Hoer's initial insight into its potential impact and role in the measurement community. Finally, the author would like to acknowledge and express his appreciation for the support and encouragement from his wife, Sadie.

The initial draft of this book was prepared during a five month tenure as guest professor, at the Technical University of Denmark, in 1986. In this context, the encouragement of Bent Knudsen and the financial support from the Otto Mønsteds Fond are gratefully acknowledged. More recently it has been used as the basis for a training course at the Singapore Institute of Standards and Industrial Research, and at the National Institute of Standards and Technology.

Glenn F. Engen
333 Sunrise Lane,
Boulder, Colo. 80302
USA

Foreword

Foreword

Occasionally one has the pleasure of reading a text by an author who has pondered over the finer intricate points of the subject in hand. This text is one such. Right through, Dr. Glenn Engen's enthusiasm comes through the pages, along with clarity of thought and a refreshing manner, that engage the readers' attention. This, of course, comes as no surprise to those who have had the pleasure of knowing Glenn.

No system in science and engineering can be successfully designed, analysed and specified, unless it is backed up by precise quantitative measurements. It is also necessary to specify error limits, precision and resolution. This is particularly difficult in microwave metrology, as more often than not, the parameter one is interested in, generally, cannot be measured directly, but has to be inferred from the measurement of other related parameters. Microwave measurements also have involved painstaking and time consuming procedures. Recently, modern 'Computer controlled and digital display' instruments have taken a lot of the drudgery out of microwave metrology. However, they do give the impression of being accurate to the last digit. As such, this book is very timely indeed, for it makes the reader aware of the pitfalls that lie in wait for the unwary experimenter, who might otherwise tend to believe everything that is digitally displayed.

This text is a must, not only for aspiring microwave metrologists, but for microwave engineers in general, as a guide to the principles of modern microwave measurement theory and practise.

Dr. S.R. Judah, M.I.E.E.E., M.I.E.E., C. Eng.
University of Hull
Hull
England.

Chapter 1

The world of microwaves

The term 'microwaves' is generally understood to refer to that portion of the electromagnetic spectrum which contains the frequencies between 300 MHz and 300 GHz and for which the corresponding wavelengths lie between one meter and one millimeter. Quite obviously the boundaries are somewhat arbitrary, but the characteristic feature is that the wavelengths tend to correspond approximately to the linear dimensions of the apparatus by which they are investigated and applied.

The basis for the prefix 'micro', which means 'small', is rooted in the development of the art. Although much of the experimental work by Hertz, which first demonstrated the existence of electromagnetic waves, was done at these wavelengths, the practical problems in their generation was such that the lower frequency portion of the electromagnetic spectrum was developed first. In addition, the engineering community was somewhat slow to recognise the practical value of these shorter wavelengths. The 1912 Radio Act, for example, 'gave' that (useless!) portion of the spectrum comprised of 'wavelengths shorter than 200 m' to the radio 'amateurs' in the belief that 'they couldn't get out of their backyards with it anyway!!'

In time, wavelengths of less than 200 m became known as 'short wave'. At the same time the adjectives used to describe the associated frequencies went from 'high' to 'very high' then 'ultra high', 'super high', 'extremely high' etc. Today the scope and variety of applications for these (useless) wavelengths have become so numerous that it represents a formidable job to even catalogue them all.

But so much for history. As already noted, 'micro' is synonymous with 'small', but in relation to what? In the present context, the wavelengths in question are only 'small' in comparison with those associated with an art that was developed at an earlier point in time. In reality, a better choice would probably be to reserve the term 'microwaves' for optical frequencies and to use 'macrowaves' for frequencies below say 300 MHz. Such a change in terminology, however, does not appear likely!

1

The microwave portion of the spectrum does enjoy a unique role in that the wavelengths tend to correspond, approximately, to the dimensions of man made objects. Although the fundamental laws of electromagnetic theory are certainly independent of frequency or wavelength, the comparative ease (or difficulty) in implementing a particular application may well depend on the ratio of wavelength to the sizes with which one conveniently works.

In any case the spectrum of electromagnetic interactions may be loosely divided into three regions according to

$$(1) \quad \lambda \gg D$$

$$(2) \quad \lambda \approx D$$

$$(3) \quad \lambda \ll D$$

where λ is the wavelength and D is some convenient dimension. Generally speaking, as one moves from one of these regions to the next, one also finds a major change in technology, although in some cases it is possible to bridge the regions by a suitable modification of the techniques in question.

The parabolic reflector, for example, which is used to focus a light beam, becomes (after a suitable increase in size!) a highly directive microwave antenna. The traditional low frequency circuit theory, in which the basic elements include 'voltage', 'current', 'impedance' etc., and with which the reader is assumed to be familiar, may with a suitable reinterpretation be applied at microwave frequencies. Indeed it is a major objective of this book to develop and exploit this application while at the same time keeping its limitations sharply in focus.

Although it is beyond the scope of this book to trace the discovery of that body of knowledge which is today known as electromagnetic theory, the development of any scientific discipline begins with the recognition of certain basic phenomena. This is then followed by attempts to quantify the observations which lead to physical laws. The work of Volta, for example, eventually found a place in Ohm's law. The discovery of Oersted is embodied in the law of Biot and Savart. Ultimately these observations, together with those of Faraday, Henry and others, were reduced to the highly condensed form which is today known as Maxwell's equations.

This step of reducing a series of observations to a physical law presupposes the ability to carefully define, replicate and quantify ones observations. In a few words, this is what the science of metrology is all about. In the words of Lord Kelvin:[1]

> *'I often say that when you can measure what you are speaking about and express it in numbers, you know something about it: but when you cannot measure it, when you cannot express it in numbers, your knowledge is of a meagre and unsatisfactory kind: it may be the beginning of knowledge, but you have scarcely, in your thoughts, advanced to the stage of <u>science</u> whatever the matter may be!'*

In many cases the role of the metrologist or scientist is substantially simplified by the introduction of a 'model'. The operation of a common flashlight, for example, may be represented on paper by the conventional symbols for a battery, resistor, (bulb) and interconnecting lines which represent the current path. In this case it is possible to make a 1:1 correspondence between the component parts of the flashlight: battery, bulb, switch and current path and the symbols on paper. It is also possible to make a 1:1 correspondence between these elements and the parameters voltage, current and resistance which are embodied in the appropriate physical law, namely that of Ohm. Moreover, in many cases the concept of a model continues to retain its usefulness, even where this 1:1 correspondence between the elements of the model, and the physical components of the system which it represents, is no longer possible. In the last analysis, a mathematical formulation of a physical law, e.g. Maxwell's equations, may also be considered a 'model' although it may also be so abstract as to inhibit if not preclude an alternative representation.

As noted, models can be extremely useful in the development of physical insights as to how nature will behave in a different situation. These mental pictures may, in turn, become the key to further development or the discovery of further relationships between the physical parameters involved. The reader is strongly cautioned,

---

[1]'Popular lectures and addresses by Sir William Thompson.' Vol.1, pp. 73-74. London, Macmillan, 1889.

however, that any conclusions which may result from this sort of mental activity should always be tested against the accepted physical laws using a carefully formulated analytical procedure.

With few (if any) exceptions, all models are incomplete and only approximately describe the phenomena they represent. In the foregoing (flashlight) model, for example, the nature of the chemical activity which is the source of the electromotive force was not included. Moreover, the internal resistance of the battery and the dependence of bulb resistance on current were both ignored. The foregoing may be summarised by noting that, 'No model is perfect; — some are more useful than others'. The choice of model to represent a particular set of circumstances is strongly dependent, not only on the questions to be asked, but also on the *accuracy* to which the answers are required. For example, if one is interested in the trajectory of a bullet over a distance of 50 m, it is *probably* appropriate to assume that the force of gravity is constant in both magnitude and direction throughout the path. On the other hand, this implied model for gravity becomes hopelessly inadequate if the target is on the other side of the world!

The most complete description of electromagnetic phenomena is provided by the field or Maxwell's equations. On the other hand, the amount of detail which is included therein is also such that, in some cases at least, the basic phenomena of primary interest are obscured. Moreover, apart from a few simple geometries, their solution may prove extremely if not prohibitively difficult. Fortunately, for a wide range of electromagnetic phenomena, the model provided by the low frequency circuit theory, already alluded to, is more than adequate for most engineering purposes. As a rule, this will be the case if the first of the above criteria ($\lambda \gg D$) is satisfied. In general, the system model which is provided by circuit theory is characterised by a small (and certainly finite!) number of discrete components or elements. The interactions among these components is via well defined current paths, which are typically provided by metallic conductors. Although electric and magnetic field phenomena and the energy storage associated therewith, are not excluded, these fields are assumed to be confined within well defined boundaries and the associated circuit elements are, of course, known as capacitance and inductance. As before, the interactions between these confined fields and the outside world is via a well defined current path, as is implicit in Kirchhoff's laws.

As a practical matter, with increasing frequency and decreasing wavelength, the problems in confining the fields, and thus limiting the interaction to the defined current paths, becomes increasingly difficult. For better or worse, the physical laws which govern the universe have no respect for the models which have been invented to describe them. As a consequence, the field phenomena play an increasingly important role in describing the interactions among the elements. This, in turn, leads to a breakdown in Kirchhoff's laws and the model contained in circuit theory.

As a general rule, the choice of model for a particular problem should include enough detail, or represent a sufficiently close approximation to the physical laws, to answer the questions of interest, at least to the desired accuracy. On the other hand, if the model is more detailed than required, its use ordinarily calls for the expenditure of unnecessary effort. Beyond this, however, the unnecessary (mathematical) details may also tend to obscure certain underlying relationships. This point is well expressed by Arguimbau[2] who writes:

'The problem is likely to be complicated by unnecessarily powerful mathematical artillery which nearly always is brought into action in the study of a new development before simple rules are appreciated.'

Although the original application of these words was in the context of noise phenomena, they apply with equal, or possibly greater force in the field of microwave metrology where the mathematical details can get completely out of hand. There are certainly problems which do call for a substantial amount of detail and some of the major contributions to the art are indebted, in no small degree, to the availability of a concise and perceptive notation as is found in matrix methods or vector analysis.

It is the major objective of this book to develop the subject of microwave metrology in a way which reflects the basic principles in as simple a manner as possible without sacrificing mathematical or analytical rigour.

---

[2]L. B. ARGUIMBAU: 'Vacuum-tube circuits', John Wiley & Sons, Inc., 1948

Chapter 2

Foundations of microwave circuit theory

As noted in Chapter 1, the low-frequency or 'conventional' circuit theory is ordinarily the best model when the wavelength is large in relation to the dimensions of the system components. With increasing frequency and decreasing wavelength, however, the ability of the circuit model to explain the observed behaviour is no longer adequate for many practical problems. In principle at least, and assuming that the phenomena of interest fall within the scope of electromagnetic theory, these limitations can be overcome by recourse to a different model, namely the Maxwell or field equations. Unfortunately, however, this model is also an extremely complicated one. Here the low frequency concepts of 'voltage' and 'current' are, to a large extent, replaced by the electric and magnetic fields associated therewith and these are then described throughout all space!
Although there are applications where this detail is useful, if not essential, there are many others where it is not and the problem is how to modify the model so it represents only the details of interest. The conventional low frequency circuit theory is one such simplification. Moreover, it can be shown that this circuit theory is a consequence of Maxwell's equations under the given conditions.

Theoretical basis for _microwave_ circuit theory

As will now be demonstrated, there is another set of conditions where a comparable simplification is possible. Indeed, the model or 'microwave circuit theory', which emerges from these considerations is in a formal sense equivalent to its lower frequency counterpart. On the other hand its validity, or scope of application, rests on a completely different set of postulates. Moreover, although these are easily stated and readily understood, their realisation in a given environment can be a never ending challenge. By contrast, the criteria for low-frequency circuit theory are, in many cases, assured by the nature of the application and require little thought on the part of the user. Indeed their 'violation', which is an implicit requirement for a radiating antenna for example, can represent a major engineering challenge.

The typical microwave system usually includes a **number of** components which are interconnected by means of a transmission line, examples of which include the rectangular waveguide and coaxial line. The essential features include one or more metallic conductors and a cross-section which is constant or independent of the position along the line. The model of interest begins with a solution to the field equations in this context. This, however, calls for concepts and related mathematical methods which are substantially beyond those required for the remainder of this text.[3] For this reason, the basic conclusions will only be stated, but then interpreted in detail so as to permit one to make an assessment of their scope of application. To be specific, it can be shown that the solution to Maxwell's equations in a uniform, isotropic and lossless transmission line, and irrespective of the nature of the cross-section (rectangular, coaxial, elliptical etc.), is given by a pair of waves propagating in opposite directions. Actually, there is a family of waves or 'modes' in each direction, but by a suitable choice of dimensions and operating frequency one can restrict the operation to a single mode. Since this is the one most commonly encountered in practise, the discussion will be confined to this case.

Unless otherwise stated, the description will be in terms of *complex amplitudes* (or *phasors* which include the *phase* as well as *magnitude* information) rather than instantaneous real quantities. In addition, rationalised MKS units will be assumed. For the implied sinusoidal excitation, the solution to Maxwell's equations includes the result that the *transverse* components ($\mathbf{E}_t$, $\mathbf{H}_t$) of the electric and magnetic fields (for the *mode* of interest) can be expressed in the form[4]

$$\mathbf{E}_t = \left[Ce^{j(\omega t - \beta z)} + De^{j(\omega t + \beta z)} \right] \cdot \mathbf{F}_t \qquad (2.1)$$

and

$$\mathbf{H}_t = \left[Ce^{j(\omega t - \beta z)} - De^{j(\omega t + \beta z)} \right] \cdot Y \cdot \mathbf{k} \times \mathbf{F}_t. \qquad (2.2)$$

---

[3]For a concise yet comprehensive treatment of this topic see D. M. KERNS, and R. W. BEATTY, 'Basic theory of waveguide junctions, and introductory microwave network analysis', Pergamon Press, 1967

[4]With the exception of the 'TEM' mode, as exists in a coaxial line, for example, there will also be a _longitudinal_ component of the electric field for 'TM' modes, and of the magnetic field for 'TE' modes. These details, however, are not necessary for the material which follows.

Here C and D are complex constants, while k is the unit vector in the z direction and which is also assumed to correspond to the axis of the transmission line. The (real) vector $\mathbf{F}_t$ lies in the transverse plane perpendicular to k (or the axial direction) and is a function only of the transverse co-ordinates which may be in rectangular, polar or some other convenient form. To be more specific, $\mathbf{F}_t$ contains the 'information' as to how $\mathbf{E}_t$ varies over the cross-section of the line. The vector operation k×$\mathbf{F}_t$ 'rotates' $\mathbf{F}_t$ by 90° in the transverse plane, thus $\mathbf{H}_t$ is perpendicular to $\mathbf{E}_t$, as expected. The parameter Y is a (real) function of the dielectric and magnetic properties (ϵ, μ) of the media which fills (or surrounds) the line. Finally, the time dependence is contained in the factor $e^{j\omega t}$ and the z dependence in $e^{\pm j\beta z}$.

The combination ($\omega t \pm \beta z$), which occurs in the exponentials, is the characteristic feature of a wave since the value of $\omega t - \beta z$ for some arbitrary $t = t_1$, $z = z_1$ is also realised at z_2, ($z_2 > z_1$) at the later time t_2 which is given by

$$t_2 = t_1 + \frac{\beta}{\omega}(z_2 - z_1)$$

The velocity v of the wave is given by

$$v = \frac{\Delta z}{\Delta t} = \frac{z_2 - z_1}{t_2 - t_1} = \frac{\omega}{\beta}$$

and this implicitly defines β since ω is, by definition, $2\pi f$ where f is the frequency. To be more specific

$$\beta = 2\pi f/v = 2\pi/\lambda$$

where λ is the wavelength. Moreover

$$\lambda = v/f = 2\pi/\beta$$

The complex exponential $e^{j(\omega t - \beta z)}$ represents a sinusoidal wave of unit amplitude in both time and position along the z axis. Thus $Ce^{j(\omega t - \beta z)}$, where $C = |C|e^{j\theta}$, represents a wave in the positive z direction of amplitude $|C|$ and whose phase for $\omega t - \beta z = 0$ is given by θ. In a similar manner, $De^{j(\omega t + \beta z)}$ represents a wave in the negative z direction.

Since the factor $e^{j\omega t}$ occurs in all terms, it may be omitted although explicitly understood. For many purposes, one's interest is often limited to a specific value of z (terminal surface) and since the origin of the co-ordinate system may be chosen to suit one's convenience, it is posssible to let $z = 0$.

Equations 2.1 and 2.2 may now be written

$$\mathbf{E}_t = [\, C + D \,] \cdot \mathbf{F}_t \tag{2.3}$$

$$\mathbf{H}_t = [\, C - D \,] \cdot Y \cdot \mathbf{k} \times \mathbf{F}_t \tag{2.4}$$

which represent the basic model for microwave circuit theory. The foregoing may thus be summarised by noting that the excitation state at the given transverse plane or 'terminal surface' is completely specified by the complex amplitudes C and D and where $\mathbf{F}_t$ and Y are determined by the geometry (or cross-section) and the properties (ϵ, μ) of the associated media. It is important to note that the field equations do not, indeed cannot, provide values for C or D. Instead they merely assure us that the solution may be expressed in the given form. In a practical application, the values for C and D are determined by the nature of the terminations at the two ends of the transmission line as will be explained in the following chapter.

Power normalisation

Although the model, as given, is quite adequate for perhaps 99% of the applications to follow, it will also prove useful to note some further refinements or extensions of it. The first of these relates to the so-called 'power normalisation'. In general, the power P may be computed by integrating the complex Poynting vector over the terminal surface of interest

$$P = \operatorname{Re} \tfrac{1}{2} \int_s \mathbf{E}_t \times \mathbf{H}_t^* \cdot \mathbf{k} \, ds \tag{2.5}$$

Substitution from eqns. 2.3 and 2.4 into this expression and recalling that the complete dependence on the transverse co-ordinates is contained in $\mathbf{F}_t$ yields, with the help of a well known vector identity

$$P = \left[\, |C|^2 - |D|^2 \,\right] \cdot Y \cdot \tfrac{1}{2} \int_s \mathbf{F}_t \cdot \mathbf{F}_t \, ds \tag{2.6}$$

At this point it is useful to note that eqn. 2.6 may also be written

$$P = \left[\, |A|^2 - |B|^2 \,\right] \left(\, w^2 \, Y \, \tfrac{1}{2} \int_s \mathbf{F}_t \cdot \mathbf{F}_t \, ds \,\right) \tag{2.7}$$

where $A = C/w$, $B = D/w$ and where w is a real number which may be chosen as convenient. The second factor in eqn. 2.7 evaluates to a constant K and, because it includes the factor w^2, it may be assigned an arbitrary value. Thus, eqn. 2.7 becomes

$$P = K\left[|A|^2 - |B|^2\right] \tag{2.8}$$

In general the factor K permits one to make an arbitrary choice of the units in which the power P and the complex amplitudes A and B are measured. However as a practical matter, and as explained in greater detail in a chapter to follow, the wave amplitudes are seldom observed directly. Instead their values are inferred from power measurements. As a consequence there is little practical need to establish the units in which A and B are measured and it is often convenient to let $K = 1$. Since the power P is usually measured in Watts, this represents an implicit assignment of the units for A and B.

Introduction of 'voltage' and 'current'

As will be developed in the chapters to follow, the foregoing solution of the field equations yields the basis for *microwave circuit theory*. Here the fundamental or basic parameters are the complex wave amplitudes A and B. This *model* applies to transmission lines of arbitrary cross-section, such as a rectangular waveguide and may be used as the basis for defining a waveguide 'voltage' and 'current' as follows.

Let V and I be defined by the equations

$$V = A + B \tag{2.9}$$

$$I = (A - B)/Z_0 \tag{2.10}$$

where for the present Z_0 is an *arbitrary* parameter. When solved for A and B these yield

$$A = \tfrac{1}{2}\left[V + IZ_0\right] \tag{2.11}$$

and

$$B = \tfrac{1}{2}\left[V - IZ_0\right] \tag{2.12}$$

Comparison with the low-frequency model

In the case of coaxial line, it will be of interest to compare this result with an alternative model, which is provided by the low-frequency transmission line theory. In the low-frequency circuit model the basic parameters are the voltage V and current I. Here a continuum of infinitesimal sections, each comprised of a series inductance and shunt capacitance, is the usual model for the transmission line. The solution to the associated differential equation may be written in the same form as eqns. 2.9 and 2.10 and where A and B are the complex amplitudes of the voltage waves in the two directions. In the low-frequency model, however, the parameter Z_0 assumes a more fundamental role. To be specific, Z_0 may be defined by the ratio V/I which obtains for an assumed wave motion in one direction only ($B = 0$) and is generally known as the *characteristic impedance*. It is important to note, however, that the numerical value which is assigned to the characteristic impedance, via this route, is a function of the cross-sectional geometry *and of the (arbitrary) conventions employed in arriving at the (low-frequency) system of units in which V and I or A and B are measured*. Although not ordinarily required, it is a simple matter to extend these conventions to the microwave circuit theory by the simple expedient of letting $K = 1/Z_0$ so that eqn. 2.8 becomes

$$P = \frac{|A|^2 - |B|^2}{Z_0} \qquad (2.13)$$

From the low-frequency circuit theory perspective, eqns. 2.11 and 2.12 are usually interpreted as providing the 'definition' for the complex wave amplitudes A and B. Moreover, in a completely formal sense, these definitions can be arbitrarily applied in the low-frequency circuit context even where the physical interpretation of representing wave amplitudes and the associated criteria which ordinarily dictates the choice of Z_0 completely breaks down.

By contrast, from the perspective of field theory, eqns. 2.9 and 2.10 may be regarded as providing the definitions of a 'generalised' voltage and current (V and I) which retain their validity even in the case of rectangular waveguide for example and where the low-frequency definitions are no longer applicable. Moreover, from this perspective, Z_0 may be regarded as an arbitrary constant and chosen to suit one's convenience. In a coaxial line (and other two- conductor or 'TEM' systems), there is a fairly compelling rationale or basis for the choice, as outlined above, which leads to a 'V' and 'I' which are equivalent to their low-frequency counterparts. In other cases the criteria are less obvious. Although a number of possible ways of defining a characteristic impedance for a waveguide have been proposed, there is no clear favourite and a common choice is to let $Z_0 = 1$. For many problems, either of the above points of view leads to the same results or conclusions. There are others, however, for example the question of how to handle departures from the assumed transmission line uniformity, where there may be a difference.

In any case it is appropriate to emphasise that the model contained in eqns. 2.3 and 2.4 is the basis of the material to follow. Although the low-frequency circuit model does provide some useful additional insights, especially in the case of a coaxial line, it also has its limits. To be more specific, both the low-frequency circuit theory and microwave circuit theory may be regarded as special cases of Maxwell's equations which, although similar in form, rest on a completely different set of postulates. These must be kept sharply in focus in efforts to assess the applicability of the model to specific problems. In keeping with this objective, these postulates will be specifically identified as the development proceeds and the analogies to low-frequency circuit theory identified as appropriate. As implicit in the above, the first of these postulates is that for a suitably restricted band of operating frequencies, the excitation state in an assumed lossless and uniform transmission line, of arbitrary but known parameters, is completely determined by the two complex amplitudes A and B. Moreover, the expression for power is given by eqn. 2.8, where K is a constant, whose value in a particular application is physically determined by the transmission line parameters (including geometry) and by the units in which P, A and B are specified.

Because the foregoing is somewhat abstract, it may be useful to consider a specific example. A waveguide of elliptical cross-section will be assumed. Moreover the excitation conditions, which have been assumed, are uniquely determined by two requirements: (1) The wave motion is in one direction only (2) The amplitude is such that the power is one Watt. The transverse electric field, whose value may be specified in the units of Volts/meter, will be a function of the position or location in the transverse plane. In order to assign a value to the wave amplitude, one must adopt some convention as to where in the cross-section (transverse plane) this is to be specified and a convenient choice might be at the ellipse centre and parallel to the minor axis. In common with other physical phenomena, one can expect that the power will be proportional to the square of the wave amplitude. Moreover, the constant of proportionality will be a function of the waveguide dimensions, frequency, dielectric and magnetic properties of the medium which fills the guide etc.

In addition, however, it will also depend on the convention which has been adopted as to where the electric field is specified. If a different convention is adopted, say for example in the direction perpendicular to the major axis and at one of the foci, a different proportionality factor will be obtained. Quite obviously, if one wants to infer a value for the power from experimental observations of the electric field amplitude, these considerations assume a substantial measure of practical importance.

In reality, however, and this is a characteristic feature of microwave metrology, the field amplitudes are seldom (if ever) observed directly. Instead, they are *usually* inferred from power measurements. For this reason the proportionality factor can remain unspecified. This point of view is adopted in the material which follows.

Chapter 3

Elementary boundary conditions

To provide a background for the material to follow, some additional characteristics of the wave solution introduced in Chapter 2 will prove useful. Returning to eqn. 2.1, it is convenient to define a scalar E whose value is proportional to the magnitude of $\mathbf{E}_t$ such that

$$E = \left[Ae^{j(\omega t - \beta z)} + Be^{j(\omega t + \beta z)} \right] \tag{3.1}$$

For a fixed position or value of z, the fields will vary sinusoidally with time in accordance with the factor $e^{j\omega t}$. However, because of the sign change in z, which occurs in the second term, the dependence on z requires some additional attention. Let eqn. 3.1 be written

$$E = \left[ae^{-j\beta z} + be^{j\beta z} \right], \tag{3.2}$$

where $a = Ae^{j\omega t}$ etc. and the time dependence has been suppressed. As noted in Chapter 2, if $b = 0$ all positions along the line are 'equivalent' in the sense that whatever occurs at $z = z_1$ also happens at $z = z_2$, $(z_2 > z_1)$ at a later time. Moreover, if the z dependence is examined at some fixed value of time, one finds a sinusoidal variation in the instantaneous amplitudes as a function of position. Consider now the case where the magnitudes of a and b are equal. Then, $b = ae^{j2\theta}$, where 2θ is the phase difference and

$$E = \left[ae^{-j\beta z} + ae^{j(\beta z + 2\theta)} \right] \tag{3.3}$$

or

$$E = ae^{j\theta} \left[e^{-j(\beta z + \theta)} + e^{j(\beta z + \theta)} \right]. \tag{3.4}$$

By use of the identity

$$\cos\phi = \tfrac{1}{2} \left[e^{j\phi} + e^{-j\phi} \right] \tag{3.5}$$

this becomes

$$E = 2ae^{j\theta}\cos(\beta z + \theta) \tag{3.6}$$

14

The electric field *amplitude* (in contrast to its instantaneous value) now varies sinusoidally along the line. By inspection it will be a maximum for $\beta z + \theta = n\pi$ and will vanish at those places where $\beta z + \theta = (2n + 1)\pi/2$. This phenomen carries the name '*standing wave*' in contrast to '*travelling wave*' since the positions of the maxima and minima do not change. Finally, in the general case, it is possible to define a parameter c such that $b = a + c$. Substitution of this in eqn. 3.2 leads to the conclusion that the general solution may be regarded as a superposition of the travelling and standing waves.

The solution to the field equations, which was given by eqns. 2.1 and 2.2, assumes that the transmission line itself is free of 'sources'. Thus, apart from a source of excitation at one (or possibly both) ends, the fields would be everywhere zero. In order to complete the solution, it is necessary to evaluate the so-called 'boundary conditions' or make an assessment of the role of the terminations which are found at the ends of the transmission line.

'Ohm's law' for passive terminations

Although the given solution represents the 'steady state' condition, there are times when a description based on the 'transient' response provides a better intuitive insight as to the behaviour. In

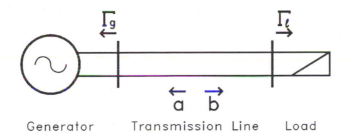

Generator   Transmission Line   Load

Fig. 3.1 Microwave source, transmission line, and termination

any case, and referring to Fig. 3.1, it will be assumed that at some arbitrary point in time the generator has been 'turned on' and a wave launched into the line. Along with the wave motion, there will be a propagation of energy which is carried by the electric and magnetic fields. Upon encountering the termination at the opposite end, one of two things can happen: the energy may be absorbed by the termination or it may be 'reflected' back towards the source. In general some of both will occur. Now, associated with the reflected energy will be a travelling wave in the opposite direction and whose amplitude may be expected to be proportional to that of the incident wave. Thus the first of the boundary conditions is given by

$$a = b\Gamma_l \tag{3.7}$$

where Γ_l is a (complex) constant of proportionality, generally known as the *reflection coefficient* and whose value is a characteristic of the termination.   The subscript 'l' (load) has been added to avoid confusion with the other uses of Γ which will follow.

At this point a convention should be noted. Ordinarily 'b' is used to represent the 'emerging' wave amplitude (away from the device of interest). In this case, this is the generator, although the immediate focus is on the termination. In addition, 'a' is used for that of the 'incident' (or incoming) wave. Unfortunately, but quite obviously however, it is not possible to completely adhere to this convention if the terminal plane is also the interface between two devices.   In general one must frequently examine the context to determine the extent of adherence to this convention (or lack thereof).

It cannot be too strongly emphasised that the foregoing does *not* constitute a derivation or proof of the given result, although hopefully the discussion has made it intuitively plausible. Instead, the 'proof' lies in the simple fact that this relationship or *model* accurately describes the observed behaviour for an extensive range of applications.

Equation 3.7 is the microwave circuit counterpart of 'Ohm's law'. Moreover, if one reverts to the low-frequency circuit model, this result could be shown to be a consequence thereof. In keeping with the above argument, however, and in the strictest sense of the word, there is no 'proof' of Ohm's law either. Its acceptance, rather, also rests on its ability to accurately model a large range of observed phenomena. A further comparison is of interest. With regard to Ohm's law, either the voltage or the current may be regarded as the stimulus or driving force and the resulting current or voltage becomes the response. Although, from a physical point of view, there may be some preference for the former, both are certainly useful.   By contrast, in the travelling wave model, there is a fairly compelling physical basis for interpreting the incident wave as the stimulus and the reflected wave as the response.

With the help of eqn. 2.8 the power delivered to the termination may be written

$$P = |b|^2 - |a|^2 \qquad (3.8)$$

where 'K' has been set equal to unity. In particular, if $a = 0$, $P = |b|^2$ and conversely. It is thus possible to define an *incident* and *reflected* power, which is given by the magnitude squared of the corresponding wave amplitudes, such that the (*net*) power P, as given by eqn. 3.8, is just their difference. After substituting the value of a from eqn. 3.7 this becomes

$$P = |b|^2 \left[1 - |\Gamma_l|^2\right] \qquad (3.9)$$

It may now be recognised that if the energy flow, or power, is indeed towards the termination, this requires $|\Gamma_l| \leq 1$ and a termination which satisfies this condition is, by definition, 'passive'. Moreover, this is ordinarily the case.[5]

Several special cases are now of interest. If $\Gamma_l = 0$, all of the (incident) energy will be absorbed; there will be no reflected wave and the termination is said to be 'matched'. (Unfortunately, however, there are a variety of other meanings which are also associated with this term.) Next, it may be that the termination fails to provide for the dissipation of energy. This requires that $P = 0$, thus from eqn. 3.8 one has $|a| = |b|$ and $|\Gamma| = 1$. This will be the case, for example, if the transmission line is terminated by a perfectly conducting plate, perpendicular to its axis, or 'short' (where the latter term has been borrowed from low-frequency circuit theory) and which requires that $\mathbf{E}_t$ (and E) vanish. By use of eqn. 3.2 and assuming that the co-ordinates have been chosen such that $z = 0$ at the short position, one has $a = -b$ and $\Gamma = -1$. The counterpart, or 'open' circuit, is one for which $\mathbf{H}_t$ vanishes, which requires $C = D$ in eqn. 2.4 and which then leads to $a = b$ or $\Gamma = 1$.

Microwave equivalent of 'Thevenin's theorem'

Next, and referring again to Fig. 3.1, with the addition of an energy source the 'termination' becomes a 'generator' (or signal

---

[5]For completeness, one may note that devices which violate this condition and which provide the basis for certain types of amplifiers have been built. Their consideration, however, is outside the scope of the present discussion.

source) whose behaviour may be usually represented by the addition of a constant term b_g to eqn. 3.7 such that it becomes

$$b = b_g + a\Gamma_g \qquad (3.10)$$

and where the subscript 'g' represents 'generator'. In particular, b is now the sum of two terms; one is a constant (b_g), the other ($a\Gamma_g$) is proportional to a. As before, no attempt will be made derive or 'prove' this result. Its acceptance rests on its ability to explain or accurately 'model' a wide range of observed phenomena. On the other hand, it should be noted that this is the most general *linear* relationship which can exist between b and a. A more general boundary condition could be postulated, e.g.

$$b = C_0 + C_1 a + C_2 a^2 + \cdots$$

but when this is done the behaviour no longer falls within the domain of *linear phenomena*.

Equation 3.10 may be regarded as the microwave circuit theory analogue of Thevenin's or Norton's theorems. As was the case with the passive termination, one could show that eqn. 3.10 is a consequence of these, but as before the justification will be allowed to rest on the arguments given. Returning to eqn. 3.10, if $\Gamma_g = 0$, b is a constant and thus independent of a. As will become apparent in due time, this represents a case of considerable practical importance, especially in measurement applications. In addition, a microwave source, for which $\Gamma_g = 0$, may be regarded as the analogue of the low-frequency 'constant voltage' or 'constant current' sources. There is, however, an important difference. Whereas an assumed constant voltage or current source also implies an *infinite* source of available energy, this is not the case for the *assumed* $\Gamma_g = 0$.

Finally, the prior statements as to the scope of the validity of eqn. 3.10 do require some qualification. In reality this may be a very *poor* model for the actual source of microwave energy. Among other things, its *frequency* may be a function of how it is terminated and this is certainly not reflected in the assumed model. However, with the help of frequency and amplitude stabilisation circuits, the source can generally be made to perform in the manner implicit in eqn. 3.10 and this will be assumed in what follows.

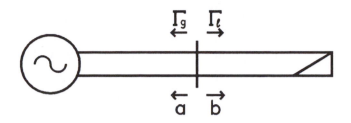

Fig. 3.2 Microwave source and passive termination

The basic ideas contained in the above material will be developed in greater detail in the Chapter 4. At this point their application to a very basic and fundamental problem will be considered. Referring to Fig. 3.2, a signal source, line and (passive) termination will be assumed. In addition, a terminal plane will be chosen at some arbitrary position as shown. Although it is common to visualise the generator and termination as connected (or separated!) by the line, it is usually more convenient, as in this case, to take the point of view that the 'generator' includes everything on one side of the terminal surface, while the 'termination' includes everything on the other side. In particular, this merely requires an appropriate adjustment of phase in b_g, Γ_g, Γ_l etc. Following this, the boundary conditions, eqns. 3.7 and 3.10 may be applied.

The elimination of a between eqns. 3.7 and 3.10 yields

$$b = \frac{b_g}{1 - \Gamma_g \Gamma_l} \tag{3.11}$$

which is the result of interest.

An alternative (intuitive) derivation

Although eqn. 3.11 is a direct consequence of the boundary conditions, an intuitive 'derivation', based on transient considerations, is also of interest. Referring again to Fig. 3.1, as an initial condition it will be assumed that the fields are everywhere zero. Then at some point in time the generator is turned 'on'. This launches a wave 'b_g' in the line, which (at some later point in time) may be observed as a wave of this amplitude at the position of the terminal surface. From here it continues to the termination where the boundary condition (eqn. 3.7) obtains. It is then reflected with an amplitude $a = b_g \Gamma_l$. The wave then returns to the generator where it is again reflected, after being 'multiplied' by Γ_g. Since the system is assumed to be linear, superposition holds and the wave amplitude in the forward direction is now $b_g + b_g \Gamma_g \Gamma_l$. Continuing with this line of argument, in the limit b is given by

$$b = b_g + b_g \Gamma_g \, \Gamma_l + b_g (\Gamma_g \Gamma_l)^2 + \cdots \qquad (3.12)$$

which may be factored to yield

$$b = b_g \left[1 + \Gamma_g \Gamma_l + (\Gamma_g \Gamma_l)^2 + \cdots \right] \qquad (3.13)$$

The second factor in this expression is an infinite series, which for $|\Gamma_g \Gamma_l| < 1$ converges to $(1 - \Gamma_g \Gamma_l)^{-1}$ as appears in eqn. 3.11.

It should probably be observed that, although this alternative treatment provides some additional intuitive insight into the operation and leads to a correct result, it should not be taken as an example of a careful analytical formulation.

Power transfer

An expression for the power transfer may be obtained by the substitution of eqn. 3.11 in eqn. 3.9

$$P = \frac{|b_g|^2}{|1 - \Gamma_g \Gamma_l|^2} \left(1 - |\Gamma_l|^2 \right) \qquad (3.14)$$

For reasons which will emerge, it will prove useful to write this expression as the product of two factors

$$P = \frac{|b_g|^2}{\left[1 - |\Gamma_g|^2 \right]} \cdot \frac{\left[1 - |\Gamma_g|^2 \right] \left[1 - |\Gamma_l|^2 \right]}{|1 - \Gamma_g \Gamma_l|^2} \qquad (3.15)$$

where the dependence on Γ_l is contained entirely in the second quotient and which will be given the label M_{gl}. Thus

$$M_{gl} = \frac{\left[1 - |\Gamma_g|^2 \right] \left[1 - |\Gamma_l|^2 \right]}{|1 - \Gamma_g \Gamma_l|^2} \qquad (3.16)$$

It is now of interest to determine the value of Γ_l for which the maximum power will be realised. As already noted, the complete dependence on Γ_l is contained in M_{gl}, thus the problem translates to a conceptually straightforward, albeit somewhat tedious, exercise in differential calculus. An alternative solution to this problem begins with the addition and subtraction of 1 to M_{gl} so that

$$M_{gl} = 1 - \frac{|1 - \Gamma_g\Gamma_l|^2 - \left(1 - |\Gamma_g|^2\right)\left(1 - |\Gamma_l|^2\right)}{|1 - \Gamma_g\Gamma_l|^2} \tag{3.17}$$

Noting that

$$|1 - \Gamma_g\Gamma_l|^2 = \left(1 - \Gamma_g\Gamma_l\right)\left(1 - \Gamma_g^*\Gamma_l^*\right) \tag{3.18}$$

the numerator of the second term in eqn. 3.17 may be expanded and after the cancellation of certain terms yields

$$\left(|\Gamma_g|^2 - \Gamma_g\Gamma_l - \Gamma_g^*\Gamma_l^* + |\Gamma_l|^2\right) = |\Gamma_l - \Gamma_g^*|^2 \tag{3.19}$$

Thus an alternative expression for M_{gl} is

$$M_{gl} = 1 - \frac{|\Gamma_l - \Gamma_g^*|^2}{|1 - \Gamma_g\Gamma_l|^2} \tag{3.20}$$

By inspection, the second term is non-negative and vanishes if $\Gamma_l = \Gamma_g^*$. This is the counterpart of the well known result from low-frequency circuit theory, namely the maximum power is realised when the load impedance equals the conjugate of the source impedance. By inspection, the maximum value of M_{gl} is unity, thus the first factor in eqn. 3.15 represents the maximum attainable, or 'available' power which will be denoted by P_g, so that eqn. 3.15 may be written

$$P = P_g M_{gl} \tag{3.21}$$

The second factor M_{gl} may be regarded as a *mismatch factor* whose physical significance is that of expressing to what extent the conditions for maximum power transfer have been satisfied. The choices of subscripts in the above, 'g' = generator, 'l' = load, reflect the dependence of the different factors on Γ_g and Γ_l.

It is of interest to observe that for $\Gamma_g = 0$, the maximum power is realised for $\Gamma_l = 0$, which corresponds to wave motion in one direction only. Although a lossless line has been assumed, this of course is an idealisation. In general, for a given rate of energy transfer, or power, the maximum transmission line efficiency is obtained if this condition ($\Gamma_l = 0$) is realised or at least closely approximated. Although this provides some basis for the common design objective of unidirectional wave motion, a more compelling motivation usually follows from other considerations. In the most elementary case of pulse modulation, for example, there is usually a substantial interest in keeping the pulse length as short as possible, since, all other things being equal, this increases the rate of information transfer. However, as may be inferred from the 'alternative' derivation of eqn. 3.11 which was given above, a substantial distortion of the modulation envelope can occur if the line length or transit time is long in comparison to the pulse width (or 'on–off' time). This, in turn, may introduce errors into the data.

Finally, the measurands or parameters required to describe the model, which have been introduced, include Γ_l, Γ_g and P_g. With regard to the latter, however, there is also a substantial, and probably more widespread, precedent for characterising the generator by the power which it will deliver to a 'matched' termination. As inspection of eqn. 3.15 reveals, this is merely $|b_g|^2$. Uniformity in terminology is *not* a characteristic feature of the microwave art!

Chapter 4

Multi-port boundary conditions

The extension of the ideas introduced in Chapter 3 to multi-port (or multi-terminal surface) devices begins with a 'two-port' as shown in Fig. 4.1. Here the 'ports' or terminal surfaces have been designated by '1' and '2', while the incident and emergent wave amplitudes are

Fig. 4.1 Passive 'two-port'

represented by a_i and b_i, respectively, and the subscripts denote the port to which they pertain. It is assumed that the two-port device is 'passive' or free of energy sources.[6] In addition to the possible reflection and/or absorption of energy, which is a characteristic of the one-port, must now be added a third option. There may be (and usually is!) a transmission of energy from one port to another. Let port 2 be terminated in a match (or *reflectionless* termination) so that $a_2 = 0$, while port 1 is connected to a signal source. Here one can expect both a reflected and a *transmitted* wave, whose amplitudes will be proportional to the incoming wave a_1. Thus one can write

$$b_1 = S_{11}a_1$$
and
$$b_2 = S_{21}a_1$$

If one now assumes the source and termination have been interchanged so that $a_1 = 0$, $a_2 \neq 0$ one can expect

$$b_1 = S_{12}a_2$$
and

---

[6] 'Active' two-ports, which contain (noise) sources, will be considered in Chapter 21.

23

$$b_2 = S_{22}a_2.$$

The proportionality factors S_{mn}, which have been introduced, carry the name 'scattering coefficients'. (The incoming wave is 'scattered' among the different ports.) Moreover the subscripts are ordered so that the first indicates the port from which the scattered wave 'emerges', while the second indicates the port at which the excitation has been applied. Thus, S_{12}, for example, yields b_1 (at terminal 1) due to an applied excitation (a_2) at terminal 2.

By hypothesis the system is linear. Therefore, one can add the above results to obtain the general case

$$b_1 = S_{11}a_1 + S_{12}a_2 \qquad (4.1)$$

$$b_2 = S_{21}a_1 + S_{22}a_2 \qquad (4.2)$$

and where the extension to devices with three or more terminals is obvious. These are the so-called 'scattering equations', which in matrix notation may be written $\mathbf{b} = \mathbf{Sa}$, where $\mathbf{b}$ and $\mathbf{a}$ are now column vectors containing the elements b_1, b_2, a_1, a_2, while the S_{mn} are contained in the 'scattering matrix' $\mathbf{S}$ such that

$$\begin{pmatrix} b_1 \\ b_2 \end{pmatrix} = \begin{pmatrix} S_{11} & S_{12} \\ S_{21} & S_{22} \end{pmatrix} \begin{pmatrix} a_1 \\ a_2 \end{pmatrix} \qquad (4.3)$$

Again, the extension to multi-port components is straightforward.

In order to obtain the relationships between the scattering matrix and the impedance and admittance matrices, which are in common use at low frequencies, the matrix counterpart of eqns. 2.11 and 2.12 may be written

$$\mathbf{a} = \tfrac{1}{2}\left[\mathbf{v} + \mathbf{Z}_0\mathbf{i}\right] \qquad (4.4)$$

$$\mathbf{b} = \tfrac{1}{2}\left[\mathbf{v} - \mathbf{Z}_0\mathbf{i}\right] \qquad (4.5)$$

where $\mathbf{Z}_0$ is a diagonal matrix whose elements are the characteristic impedances, which have been assigned at the different terminal surfaces, while the remaining terms are column vectors.

Let $\mathbf{v} = \mathbf{Zi}$, where $\mathbf{Z}$ is the impedance matrix and $\mathbf{b} = \mathbf{Sa}$. The substitution of these into eqns. 4.4 and 4.5 yields, with the help of a little matrix algebra,

$$\mathbf{Z} = \left[1 - \mathbf{S}\right]^{-1}\left[1 + \mathbf{S}\right]\mathbf{Z}_0 = \mathbf{Y}^{-1} \qquad (4.6)$$

where $\mathbf{Y}$ is the admittance matrix.

Now

$$(1 - \mathbf{S})^{-1}(1 + \mathbf{S}) = (1 - \mathbf{S})^{-1}(1 + \mathbf{S})(1 - \mathbf{S})(1 - \mathbf{S})^{-1}$$

$$= (1 - \mathbf{S})^{-1}(1 - \mathbf{S})(1 + \mathbf{S})(1 - \mathbf{S})^{-1}$$

$$= (1 + \mathbf{S})(1 - \mathbf{S})^{-1}$$

thus $(1 - \mathbf{S})^{-1}$ and $(1 + \mathbf{S})$ commute, so that one can write

$$\mathbf{ZZ}_0^{-1} = \left[1 - \mathbf{S}\right]^{-1}\left[1 + \mathbf{S}\right] = \left[1 + \mathbf{S}\right]\left[1 - \mathbf{S}\right]^{-1}. \qquad (4.7)$$

When solved for $\mathbf{S}$ this yields

$$\mathbf{S} = \left[\mathbf{ZZ}_0^{-1} - 1\right]\left[\mathbf{ZZ}_0^{-1} + 1\right]^{-1} = \left[\mathbf{ZZ}_0^{-1} + 1\right]^{-1}\left[\mathbf{ZZ}_0^{-1} - 1\right] \qquad (4.8)$$

In addition to the scattering matrix description, the 'cascading matrix' $\mathbf{R}$ is also of interest, particularly in the case of two–port networks. This is defined by the requirement

$$\begin{pmatrix} b_1 \\ a_1 \end{pmatrix} = \begin{pmatrix} R_{11} & R_{12} \\ R_{21} & R_{22} \end{pmatrix} \begin{pmatrix} a_2 \\ b_2 \end{pmatrix} \qquad (4.9)$$

which thus provides b_1 and a_1 as functions of a_2 and b_2. The motivation for this definition may be recognised with the help of Fig. 4.2, where a second two–port has been added. Here, although the

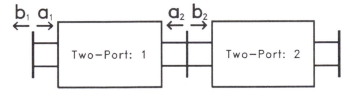

Fig. 4.2 Two 'two-ports' connected in <u>cascade</u>

definitions for b_2 and a_2 are the same as in Fig. 4.1, a_2 is now the *emerging* wave from the *second* two-port and b_2 is the incident wave. If the roles of a_2 and b_2 are reversed to reflect this, it may be recognised that eqn. 4.9 provides the relationships between the excitation conditions at terminal 1 and those at terminal 2 in such a format that the cascading matrix of two two-port networks in *cascade* is merely the matrix product of those for the individual ones.

The individual elements in **R** may be obtained by solving eqns. 4.1 and 4.2 for b_1 and a_1 , as functions of b_2 and a_2, which leads to

$$R_{11} = (S_{12}S_{21} - S_{11}S_{22})/S_{21}$$

$$R_{12} = S_{11}/S_{21}$$

$$R_{21} = -S_{22}/S_{21}$$

$$R_{22} = 1/S_{21}$$

(4.10)

while the inverse relationships are

$$S_{11} = R_{12}/R_{22}$$

$$S_{12} = (R_{11}R_{22} - R_{12}R_{21})/R_{22}$$

$$S_{21} = 1/R_{22}$$

$$S_{22} = -R_{21}/R_{22}$$

(4.11)

Finally, as will become apparent in the material to follow, it is frequently more convenient to describe the two-port operation by yet another set of parameters, A, B and C (not to be confused with their prior use in an earlier chapter) whose definitions are as follows:

$$A = (S_{12}S_{21} - S_{11}S_{22}) = R_{11}/R_{22} \qquad (4.12)$$

$$B = S_{11} = R_{12}/R_{22} \qquad (4.13)$$

$$C = -S_{22} = R_{21}/R_{22} \qquad (4.14)$$

Losslessness and realisability

By hypothesis, the two-port is *passive* which requires that the power input is non-negative. Thus the power carried by the emerging waves cannot exceed that for the incoming waves. Provided that the same choices of 'characteristic impedance' and units for power have been made at both terminals, one has

$$|b_1|^2 + |b_2|^2 \leq |a_1|^2 + |a_2|^2 \qquad (4.15)$$

where the equality is realised only in the hypothetical case of a 'lossless' network, while the inequality implies that the network is 'realisable' in the sense that its realisation does not call for the violation of energy conservation. Considering first the lossless case, one has after substitution from eqns. 4.1 and 4.2

$$|S_{11}a_1 + S_{12}a_2|^2 + |S_{21}a_1 + S_{22}a_2|^2 = |a_1|^2 + |a_2|^2 \qquad (4.16)$$

Since this holds for all choices of a_1 and a_2, setting them in turn equal to zero yields

$$|S_{11}|^2 + |S_{21}|^2 = 1 \qquad (4.17)$$

$$|S_{12}|^2 + |S_{22}|^2 = 1 \qquad (4.18)$$

Equation 4.16 may next be expanded to obtain

$$|S_{11}a_1|^2 + |S_{12}a_2|^2 + S_{11}S_{12}^*a_1a_2^* + S_{11}^*S_{12}a_1^*a_2 + |S_{21}a_1|^2$$

$$+ |S_{22}a_2|^2 + S_{21}S_{22}^*a_1a_2^* + S_{21}^*S_{22}a_1^*a_2 = |a_1|^2 + |a_2|^2 \qquad (4.19)$$

which with the help of eqns. 4.17 and 4.18 becomes

$$(S_{11}S_{12}^* + S_{21}S_{22}^*)a_1a_2^* + (S_{11}^*S_{12} + S_{21}^*S_{22})a_1^*a_2 = 0. \qquad (4.20)$$

Since the terms in eqn. 4.20 are conjugates of each other, this can only hold for arbitrary choices of a_1 and a_2 if

$$S_{11}S_{12}^* + S_{21}S_{22}^* = 0 \qquad (4.21)$$

Several immediate consequences of eqns. 4.17, 4.18 and 4.21 are of interest. If reciprocity obtains, as outlined in the section to follow, $S_{12} = S_{21}$. Referring to eqns. 4.17 and 4.18, it is then easily recognised that

$$|S_{11}| = |S_{22}| \qquad (4.22)$$

The same conclusion is also an immediate consequence of eqn. 4.21 but which also requires

$$\text{Arg}(S_{11}) - \text{Arg}(S_{12}) = \text{Arg}(S_{21}) - \text{Arg}(S_{22}) + (2n + 1)\pi \qquad (4.23)$$

There are thus three constraints on the six parameters implicit in the complex S_{11}, S_{12} and S_{22} and, as might have been expected from low-frequency considerations, the number of parameters required to characterise the two-port have been reduced from six to three.

Turning next to the question of evaluating the constraints on the matrix, S, which are implied by the inequality in eqn. 4.15, a simple extension of the arguments given above yields

$$|S_{11}|^2 + |S_{21}|^2 < 1 \qquad (4.24)$$

$$|S_{12}|^2 + |S_{22}|^2 < 1 \qquad (4.25)$$

The generalisation of eqn. 4.21 is beyond the scope of this text,[7] but the result of primary interest will be outlined. In particular, given the same power and impedance 'normalisations' which were assumed in the context of eqn. 4.15, one can show that the matrix

$$\left[1 - S^*S\right]$$

is 'positive definite'. In the above expression, the matrix S^* is the transposed conjugate (or 'Hermitian' conjugate) of S, while '1' is the unit matrix. Moreover, this result holds for the general multi-port case. By definition, a 'positive definite' matrix is one for which the determinants of all of the principal minors are positive. In the case of immediate interest, this first requires the diagonal elements to be positive which yields eqns. 4.24 and 4.25. The requirement that the determinant be positive may be written

$$\left[1 - |S_{11}|^2 - |S_{21}|^2\right]\left[1 - |S_{12}|^2 - |S_{22}|^2\right]$$
$$- |S_{11}S_{12}^* + S_{21}S_{22}^*|^2 > 0 \qquad (4.26)$$

In practise, an alternative form of this result has proven to be more easily applied to practical problems. This will now be derived. First, by means of eqns. 4.12, 4.13 and 4.14 and after expansion,

---

[7]For additional details see: D. M. KERNS, and R. W. BEATTY: 'Basic theory of waveguide junctions, and introductory microwave network analysis', Pergamon Press, 1967

cancellation and recombination, it is possible to write eqn. 4.26 in the alternative form

$$1 + |A|^2 - |B|^2 - |C|^2 - 2|A - BC| > 0 \qquad (4.27)$$

where $S_{12} = S_{21}$ has been assumed. This result is now multiplied by $(1 - |C|^2)$.

Next, after adding and subtracting $|A - BC|^2$ to complete the squares, one obtains

$$\left[1 - |C|^2 - |A - BC|\right]^2 > |B - AC^*|^2. \qquad (4.28)$$

With the help of eqn. 4.25 it will be recognised that the left side of eqn. 4.28 is positive, even without being squared. Thus, after taking the square root of both sides, one has

$$1 - |C|^2 - |A - BC| > |B - AC^*| > 0 \qquad (4.29)$$

which is the result of primary interest. It will next be shown that $|A| \leq 1$. Starting with $|S_{11}S_{12}^* + S_{21}S_{22}^*| \geq 0$ and expanding one has

$$|S_{11}S_{12}|^2 + |S_{21}S_{22}|^2 + S_{11}S_{22}S_{12}^*S_{21}^* + S_{11}^*S_{22}^*S_{12}S_{21} \geq 0 \qquad (4.30)$$

After adding and subtracting $|S_{12}S_{21}|^2$ and $|S_{11}S_{22}|^2$ this may also be written

$$\left[|S_{11}|^2 + |S_{21}|^2\right]\left[|S_{12}|^2 + |S_{22}|^2\right]$$
$$- |S_{12}S_{21} - S_{11}S_{22}|^2 \geq 0 \qquad (4.31)$$

By use of eqns. 4.24 and 4.25 the factor on the left may be replaced by unity without violating the inequality and recalling the definition of A from eqn. 4.12 one has

$$1 - |A|^2 \geq 0 \qquad (4.32)$$

as was to be proved.

At this point it is also possible to express the lossless criteria in an alternative form. In particular

$$1 - |C|^2 - |A - BC| = 0 \qquad (4.33)$$

and

$$B - AC^* = 0 \qquad (4.34)$$

Moreover, eqn. 4.22 becomes

$$|B| = |C| \qquad (4.35)$$

which in combination with (4.34) yields

$$|A| = 1 \qquad (4.36)$$

Finally, since eqn. 4.27 is symmetric in B and C, the results of eqns. 4.28, 4.29, 4.33 and 4.34, with B and C interchanged, are also obtained. An application of this will be made in Chapter 5.

Reciprocity

In low-frequency circuit theory the condition $Z_{mn} = Z_{nm}$ generally holds, and the similar result $S_{mn} = S_{nm}$ may be usually assumed in microwave circuit theory, although the conditions are somewhat more restrictive. Provided that the multi-port of interest contains only materials whose permittivity, permeability and conductivity may be represented by *symmetric tensors* (which includes scalars), the Lorentz reciprocity theorem holds. From this one can show[8] that $Z_{mn} = Z_{nm}$, or in matrix form $Z = \tilde{Z}$, where '~' represents the matrix transpose. Given this result, one has from eqn. 4.6,

$$(1 - S)^{-1}(1 + S)Z_0 = \tilde{Z}_0(1 + \tilde{S})(1 - \tilde{S})^{-1} \qquad (4.37)$$

But Z_0 is *diagonal*, thus $Z_0 = \tilde{Z}_0$ and after cross multiplication (4.37) may be written

$$(1 + S)Z_0(1 - \tilde{S}) = (1 - S)Z_0(1 + \tilde{S}) \qquad (4.38)$$

which after expansion and cancellation becomes

$$SZ_0 = Z_0\tilde{S} \qquad (4.39)$$

---

[8]For a more complete discussion see, D. M. KERNS: 'Basis of the application of network equations to waveguide problems', *J. Res. Nat. Bur. Stand.* vol. 42, pp. 515-540, May, 1949

Thus $S_{mn} = S_{nm}$ provided that $Z_{ii} = Z_{jj}$, which assumes that the same choices of characteristic impedance have been made for all terminals or ports. As a rule there is little reason to do otherwise, but given a two-port with coaxial terminals of different cross-section one may want to use the low-frequency conventions at both terminals and the reciprocity relationship becomes

$$S_{12}Z_{02} = S_{21}Z_{01} \qquad (4.40)$$

where the characteristic Z_{01} and Z_{02} obtain at terminals 1 and 2 respectively.

Finally, with regard to the assumed properties of the media which is included in the multi-port structure, it should be noted that these are well satisfied for a large range of materials. Thus reciprocity ordinarily holds. On the other hand, these criteria are not satisfied by magnetised ferrites, for example, and an important family of nonreciprocal microwave devices exist which exploit these phenomena. These include circulators and isolators.

Chapter 5

Elementary two-port applications

The theory developed in the preceding chapters will now be illustrated in some elementary two-port applications which include,

(1) Adjust load or generator 'impedance'[9]

(2) Isolate generator from changes in load impedance

(3) Provide predetermined changes in power level

(4) Provide transition between two different transmission types, e.g. waveguide to coaxial line.

These will be described in the given order.

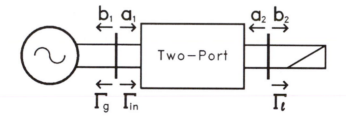

Fig. 5.1 Basic circuit for two-port applications

First, and referring to Fig. 5.1, the reflection coefficient, Γ_{in}, which obtains at terminal 1 'looking into' the two-port, will be obtained as a function of that for the load or termination, Γ_l. Referring to (4.9) one has

$$b_1 = R_{11}a_2 + R_{12}b_2 \tag{5.1}$$

$$a_1 = R_{21}a_2 + R_{22}b_2 \tag{5.2}$$

---

[9]In common with existing practice, the term *impedance*, as used in this book, includes reflection coefficient and related concepts.

Recalling that Γ_{in} and Γ_l are the ratios b_1/a_1 and a_2/b_2, respectively, the ratio of eqns. 5.1 to 5.2 yields, after division by $R_{22}b_2$,

$$\Gamma_{in} = \frac{A\Gamma_l + B}{C\Gamma_l + 1} \qquad (5.3)$$

where eqns. 4.12 $\cdots$ 4.14 have been used. The relationship between Γ_{in} and Γ_l is thus given by the ratio of two linear functions of Γ_l. This is our first example of the 'bi-linear' or 'linear fractional' transformation which plays an extremely important role in microwave metrology and about which much more will be said in a chapter to follow. In terms of the scattering coefficients this may be written

$$\Gamma_{in} = S_{11} + \frac{S_{12}S_{21}\Gamma_l}{1 - S_{22}\Gamma_l} \qquad (5.4)$$

or alternatively

$$\Gamma_{in} = \frac{(S_{12}S_{21} - S_{11}S_{22})\Gamma_l + S_{11}}{-S_{22}\Gamma_l + 1} \qquad (5.5)$$

Equation 5.4 lends itself to a convenient physical interpretation which closely follows the 'alternative' derivation of (3.11) found in Chapter 3. In particular, the reflected wave is the sum of two components. The first, which is due to S_{11}, is independent of Γ_l and is obtained even when $\Gamma_l = 0$. The second component may be identified with the reflection of the transmitted wave by Γ_l. To be more specific, the transmitted wave suffers an 'attenuation' S_{21} during transmission. Upon reflection it is then 'multiplied' by the factor Γ_l and finally 'attenuated' again by the factor S_{12} during the return trip. The factor $(1-S_{22}\Gamma_l)^{-1}$ takes care of the 'multiple' reflections between Γ_l and S_{22} as explained in Chapter 3.

Adjustment of load impedance (or tuning transformer)

In terms of the above formulation a (waveguide) tuning transformer (or 'tuner') is a two-port of adjustable parameters such that, for an arbitrary value of Γ_l, another arbitrary (but usually zero!) value of Γ_{in} can be realised. As a rule, it is desirable to minimise the dissipation in the tuner itself, thus these are usually low-loss devices. This, in turn, implies a large range of potential adjustment. Although rarely a practical problem, the loss characteristics can limit one's ability to realise values of $|\Gamma_{in}|$ close to unity since in the limit this implies the total absence of dissipation. In addition there may be a problem in achieving $\Gamma_{in} = 0$ if $|\Gamma_l| \to 1$. Referring to eqn. 5.3, Γ_{in} vanishes when $A\Gamma_l + B = 0$. For a low loss device one has $|A| \approx 1$, thus $|A\Gamma_l| \approx 1$, which then requires $|B| \approx 1$ and which may not be attainable due to the inevitable losses in the device. In general it is convenient to visualise the operation as one which provides a second reflection 'B' of the correct magnitude and phase to cancel that due to $A\Gamma_l$. As a rule, unfortunately, these devices tend to be frequency sensitive and, with the existing emphasis on broadband operation, there is an increasing effort to either eliminate unwanted reflections at their 'source', during the course of system development or alternatively, and particularly in metrology applications, to compensate for their existence by the use of a more complete model.

Adjustment of generator parameters

Referring again to Fig. 5.1, the extension of these results to a signal source calls for obtaining b_2 as a function of a_2 subject to the generator boundary condition, as given by eqn. 3.10

$$a_1 = b_g + b_1 \Gamma_g \tag{5.6}$$

Substitution of the values of a_1 and b_1, which are given by eqns. 5.1 and 5.2, into eqn. 5.6 yields

$$R_{21}a_2 + R_{22}b_2 = b_g + (R_{11}a_2 + R_{12}b_2)\Gamma_g \tag{5.7}$$

which may now be solved for b_2 to yield

$$b_2 = \frac{b_g}{R_{22} - R_{12}\Gamma_g} + \frac{R_{11}\Gamma_g - R_{21}}{R_{22} - R_{12}\Gamma_g} a_2 \tag{5.8}$$

and which in scattering notation may be written

$$b_2 = \frac{b_g S_{21}}{1 - S_{11}\Gamma_g} + \frac{(S_{12}S_{21} - S_{11}S_{22})\Gamma_g + S_{22}}{1 - S_{11}\Gamma_g} a_2 \tag{5.9}$$

The factor which multiplies a_2 in eqn. 5.9 is the 'generator reflection coefficient' which obtains at terminal 2. As might have been expected from linearity considerations, this is equivalent to eqn. 5.5 but where the interchange between S_{11} and S_{22} results from the 'termination' being connected to terminal 1 rather than 2. The first term on the left side in eqn. 5.9 is the new value of 'b_g' and where the modification tends to be dominated by the factor S_{21} since the magnitudes of both S_{11} and Γ_g are ordinarily small. As before, one can use a tuning transformer to make $\Gamma_g = 0$, which is of substantial importance in certain measurement applications.

Isolation

Although more elegant methods which involve the use of feedback are usually preferred, it may be necessary to 'isolate' the generator from changes in load impedance by elementary techniques. From eqn. 5.4, one procedure for doing this is to make $|S_{12}S_{21}|$ small. However if reciprocity obtains, this is also accompanied by a reduction in '$|b_g|$'. In spite of this, in an earlier era, it was frequently necessary to 'pad' or insert an attenuator (typically 20 dB, or more, for which $|S_{12}S_{21}| = 0.01$) at the generator output in order to achieve the desired performance. This also provided a 'source impedance' which was largely dependent only on the two-port parameter, S_{22}. At the same time, 99% of the potentially available signal power was 'wasted'. The advent of the 'isolator' provided a substantial measure of relief for this problem. This is a nonreciprocal device for which, ideally, $S_{12} = 0$, $S_{21} = 1$. In practise, values of $|S_{12}|^2$ as small as 0.01-0.001, i.e. 20-30 dB of isolation are now common. It should be observed, however, that the 'isolation' actually provided by a 20 dB isolator is only the equivalent of that for a 10 dB attenuator 'pad'. The reason for this may be recognised by inspection of eqn. 5.4. In particular, the isolator seeks only to reduce the value of $|S_{12}|$ while the attenuator also reduces $|S_{21}|$. At the same time, ideally at least, the isolator leaves $|b_g|$ unaltered.

Attenuation

Perhaps the most extensive application of a two-port device is as an attenuator. Referring to Fig. 5.2 it is convenient to postulate a

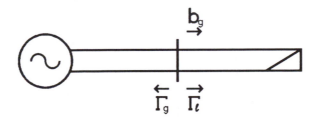

Fig. 5.2 Matched generator and termination

'matched' system such that $\Gamma_l = \Gamma_g = 0$.   Next, let the two-port be inserted between the generator and load (as shown in Fig. 5.1). The resulting reduction in power, expressed as its ratio to the prior value, is, by definition, the attenuation. The expression for attenuation, in terms of the scattering parameters is perhaps most easily obtained as an application of eqn. 5.9, where the insertion of the attenuator may be regarded as a modification of the generator parameters. In the first case, and in the absence of load reflections, the power is given by $|b_g|^2$. In the second, and noting that by hypothesis $\Gamma_g = 0$, the power is $|S_{21}|^2 |b_g|^2$. Thus the ratio is $|S_{21}|^2$. It should be noted that no assumptions as to the values of S_{11} or S_{22} have been made, although $S_{11} = S_{22} = 0$ is the usual design objective. This means that *attenuation* may result either from energy *dissipation* within the device itself *or* from *reflecting* the energy back towards the source. Denoting the initial and final values of power by P_i and P_f, respectively, the foregoing provides the basis for the usual definition

$$\text{Attenuation} = -10\log(P_f/P_i) = -20\log_{10}|S_{21}| \qquad (5.10)$$

which, as already noted, gives the reduction in power that results when the device is inserted in a *matched* system. The more general problem is far more complicated and will be deferred to a later chapter. As a rule, the term 'attenuation' is used in conjunction with two-port devices whose primary function is that of providing an *approximately* known reduction in power level, e.g. 20 dB, 40 dB etc. when inserted into a (nominally) matched system and where the errors due to deviations from the matched condition are ignored.

(Adapter) efficiency

The loss characteristics of another class of two-port devices, namely 'adapters' or transition devices from one waveguide type to another, are also of substantial interest. In this case, however, the loss or energy dissipation is usually an unwanted, but unavoidable, departure from the design objectives of providing a reflection free and dissipation free transition. In this context, the *efficiency* η, whose deviation from the value unity is a measure of the dissipation which occurs therein, is more often the parameter of interest. This, by definition, is the ratio of the (net) power output to the power input and, as will become apparent, this *does* depend on one of the system parameters, namely the load impedance. With the help of eqn. 3.8 the efficiency is given by

$$\frac{P_{out}}{P_{in}} = \frac{|b_2|^2 - |a_2|^2}{|a_1|^2 - |b_1|^2} = \frac{|b_2|^2 \left(1 - |\Gamma_l|^2\right)}{|a_1|^2 \left(1 - |\Gamma_{in}|^2\right)} \qquad (5.11)$$

Moreover, from eqn. 4.2 one has

$$\frac{b_2}{a_1} = \frac{S_{21}}{1 - S_{22}\Gamma_l} \qquad (5.12)$$

which in combination with eqn. 5.5 may be substituted in eqn. 5.11 to obtain the result of interest

$$\eta_{21} = \frac{|S_{21}|^2 \left(1 - |\Gamma_l|^2\right)}{|1 - S_{22}\Gamma_l|^2 - |(S_{12}S_{21} - S_{11}S_{22})\Gamma_l + S_{11}|^2} \qquad (5.13)$$

where the subscript '21' has been added to indicate that the energy flow is from port 1 to port 2. For a matched termination ($\Gamma_l = 0$) this becomes

$$\eta_{21} = \frac{|S_{21}|^2}{1 - |S_{11}|^2} . \qquad (5.14)$$

For certain applications the *maximum* value of efficiency is of interest. This will be considered in Chapter 15.

Returning again to Fig. 5.1, the ratio of *available* power at terminal 2 to that at terminal 1 is also of interest. For terminal 1, the result obtained in Chapter 3 eqn. 3.16 is given by

$$P_{g1} = \frac{|b_g|^2}{\left(1 - |\Gamma_g|^2\right)} \qquad (5.15)$$

while at terminal 2 one has, with the help of eqn. 5.9,

$$P_{g2} = \frac{|S_{21}b_g|^2}{|1 - S_{11}\Gamma_g|^2} \times$$

$$\frac{|1 - S_{11}\Gamma_g|^2}{|1 - S_{11}\Gamma_g|^2 - |(S_{12}S_{21} - S_{11}S_{22})\Gamma_g + S_{22}|^2} \tag{5.16}$$

The ratio of (5.16) to (5.15) now yields

$$\frac{P_{g2}}{P_{g1}} = \frac{|S_{21}|^2 \left(1 - |\Gamma_g|^2\right)}{|1 - S_{11}\Gamma_g|^2 - |(S_{12}S_{21} - S_{11}S_{22})\Gamma_g + S_{22}|^2} \tag{5.17}$$

Comparison of this result with (5.13) reveals that, except for an interchange of S_{11} and S_{22} and the replacement of Γ_l by Γ_g, they are equivalent.

An application of 'realisability'

The 'realisability' criteria which were developed in the preceding chapter will now be applied to the following problem. As noted, one of the conditions on the scattering parameters for a lossless two-port may be written $|S_{11}| = |S_{22}|$ (or $|B| = |C|$). For a low loss (or high efficiency) two-port it may be expected that this condition is approximately satisfied and an evaluation of this approximation is of value in estimating the error limits for certain types of measurement.

Provided that reciprocity holds, the counterpart of (5.14), for energy transmission from port 2 to port 1, may be written in the alternative form

$$\eta_{12} = \frac{|A - BC|}{1 - |C|^2} \tag{5.18}$$

Returning to (4.29), after division by $\left(1 - |C|^2\right)$ this may be written

$$1 - \eta_{12} > \frac{|B - AC^*|}{1 - |C|^2} > 0 \tag{5.19}$$

Thus as $\eta \to 1$ the term $|B - AC^*|$ is 'squeezed' to zero. As previously noted, in the limit $|A| = 1$ and $|B| = |C|$. If $|B|$ and $|C|$ are 'large' in relation to $1 - \eta_{12}$ the arguments of B and AC^* must differ by approximately π in order to keep $|B - AC^*|$ small. On the other hand, if $|B|$ and $|C|$ are small in relation to $(1 - \eta_{12})$, the inequality exercises little or no control over the phases or their relative magnitudes. The foregoing may be put in a more definitive form by noting

$$B - AC^* = B\left[1 - |C|^2\right] - C^*\left[A - BC\right] \tag{5.20}$$

which may be substituted in eqn. 5.19 to yield

$$1 - \eta_{12} > \left|\,|B|\,e^{j\theta} + \eta_{12}|C|\,\right| \tag{5.21}$$

where θ is a function of the arguments of A, B and C, but which is not needed in what follows. The constraint imposed on the magnitude of S_{11} (or B) for a given η_{12} and S_{22} (or C) may now be obtained from

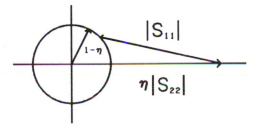

Fig. 5.3 Constraint on $|S_{11}|$

inspection of Fig. 5.3. In particular, S_{11} and θ must be so chosen that the sum $\left[\,|S_{22}|\,\eta_{12} + |S_{11}|\,e^{j\theta}\right]$ lies on or within the circle of radius $(1 - \eta_{12})$. For $|S_{22}|\,\eta_{12} > (1 - \eta_{12})$, the limits on $|S_{11}|$ are given by

$$|S_{22}|\,\eta_{12} - (1 - \eta_{12}) \leq |S_{11}| \leq |S_{22}|\,\eta_{12} + (1 - \eta_{12}) \tag{5.22}$$

For $|S_{22}|\,\eta_{12} < (1 - \eta_{12})$ the lower limit for $|S_{11}|$ is zero. In combination with eqn. 5.22, this is the result of interest.

Chapter 6

An elementary directional coupler application

Amplitude stabilisation
and multi-channel isolation

As noted in Chapter 3, the generator or source model which is implicit in eqn. 3.10 and which is repeated below,

$$b = b_g + a\Gamma_g \qquad (6.1)$$

may in reality be a poor representation of a given source of microwave energy. On the other hand, with appropriate techniques it is possible, in principle at least, to achieve any desired degree of conformity to the postulated model. In general there are two problems: frequency instability and amplitude instability. In its simplest form, frequency instability may be corrected by sampling the signal of interest, comparing its frequency against some reference and the introduction of appropriate feedback as required to achieve the desired result. This chapter will attempt to develop an understanding of amplitude stabilisation and consider some related topics.

The technique of interest usually begins with the addition of a 'directional coupler' to the output of the generator as shown in Fig. 6.1 and where the different terminals have been numbered as

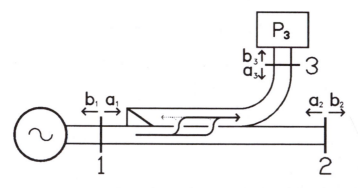

Fig. 6.1 The addition of a directional coupler provides a useful source monitor.

40

shown. A directional coupler is in reality a *four port* device. However a common arrangement includes, as assumed here, an internal termination in one of the arms such that a three-port results. This technique enjoys the dubious distinction of including what is perhaps the most widely misunderstood circuit to be found in the microwave art. In reality, there are at least three possible and distinct applications and much of the confusion appears to result from a failure to keep them that way. In an effort to clarify matters, each of these will be considered in detail. First, however, a brief description of the directional coupler will be given.

Directional coupler

Although directional couplers come in a large variety of configurations, a simplified explanation, based on rectangular waveguide, will be given with the help of Fig. 6.1. Here the 'secondary' waveguide (the one on top) is coupled to the primary one by means of a pair of holes which are separated in distance by a quarter of the guide wavelength. Individually, these provide two separate sources of excitation for the secondary guide. As a result of their separation in distance however and because these 'sources' are actually driven by the wave in the primary guide, the phase of the second (from the left) will lag the first by 90°. For this reason, the secondary waves will tend to add in phase in the forward direction and cancel in the reverse. (Note that the 'distance' to a point in the secondary guide, and to the left of the coupling holes, is a half wavelength farther from the source for coupling via the second hole as compared with the first. By contrast, for points in the secondary guide and to the right of the holes, the distances are the same.) From symmetry considerations it is apparent that the same argument will hold for wave motion in the reverse direction, thus port 3 will be more strongly coupled to a forward wave than to one in the reverse direction. For the reverse wave, the associated energy will be dissipated primarily in the internal termination. In any case, a directional coupler provides *directive coupling* to the wave motion in the main transmission line and one of the figures of merit is given by the ratio $|S_{31}/S_{32}|$ which is usually expressed

$$Directivity = 20\log_{10}|S_{31}/S_{32}|$$

In practise directivities of 20-40 dB are common, although the larger values are usually found only in waveguide systems.

The scattering equations for Fig. 6.1 may be written

$$b_1 = S_{11}a_1 + S_{12}a_2 + S_{13}a_3 \tag{6.2}$$

$$b_2 = S_{21}a_1 + S_{22}a_2 + S_{23}a_3 \tag{6.3}$$

$$b_3 = S_{31}a_1 + S_{32}a_2 + S_{33}a_3 \tag{6.4}$$

where the added feature includes the signal b_3 which is an indicator, in some sense, of the excitation present in the coupler.

Applications

As noted above, there at least three distinct applications or modes of operation *which are characterised by the ways in which the signal b_3 is used.* These include

(1) **Ignore it.** *(Yes, that is correct!!)*

(2) Observe or 'measure' it

(3) Incorporate it in a feedback loop
such that $|b_3|$ is constant

For convenience these will be considered individually in the order 1, 3, 2.

'Ignore it'

In this mode of operation the signal b_3 is (YES), *ignored!* As a consequence, the operation reverts to that of a two-port which has been added to the generator output and whose characteristics are now modified as described in the section 'Modification of signal source' in Chapter 5. In particular, there will be a change in both available power and source impedance, as dictated by the coupler characteristics, although as a rule these changes should not be large. At this point the reader may *(hopefully!)* question the motivation for adding the coupler and then failing to make any use of its response. Good question! The author has no answer, yet it is surprising to observe how frequently the considerations which apply here are interspersed and confused with those from the others. In any case, this 'mode of operation' has been identified as such only to emphasise that *this is not ordinarily desired.*

Amplitude stabilisation

Perhaps the most common and certainly the most useful application is that of amplitude or source stabilisation. In addition to a detector on arm 3, which provides $|b_3|^2$ (which, by definition, is

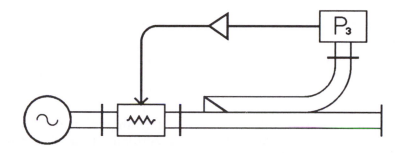

Fig. 6.2 Addition of feedback to provide amplitude stabilization.

also P_3), this calls for the addition of a feedback loop, as shown in Fig. 6.2, either in combination with an electrically adjustable attenuator or by other means (a modulator) such that P_3 is held constant. As noted above, the signal provided by P_3 will be primarily a measure of the forward wave amplitude, which appears as b_2 at port 2. Ideally, it is this component of the excitation which is held constant by the feedback loop. As noted in Chapter 3, this is the characteristic feature of a source for which $\Gamma_g = 0$. Thus it may be recognised that in addition to compensating for potential source instability, the technique also provides the equivalent of a *matched generator*.

In order to put this operation in more definitive form, one begins by eliminating a_1 between eqns. 6.3 and 6.4 and solving for b_2. After making the substitution, $a_3/b_3 = \Gamma_d$, and where Γ_d is the reflection coefficient of the detector on port 3, one has

$$b_2 = b_3 \left\{ \frac{S_{21}}{S_{31}} + \Gamma_d \left[S_{23} - \frac{S_{21}S_{33}}{S_{31}} \right] \right\} + a_2 \left(S_{22} - \frac{S_{21}S_{32}}{S_{31}} \right)$$

$$(6.5)$$

The factor which multiplies b_3 in this expression is a function of the coupler and detector parameters (and hopefully constant!) while the magnitude of b_3 is held constant by the feedback action.   The comparison of this result with eqn. 6.1 indicates that the factor Γ_{gs}, which multiplies a_2 in eqn. 6.5,

$$\Gamma_{gs} = S_{22} - \frac{S_{21}S_{32}}{S_{31}} \tag{6.6}$$

is now the 'equivalent' reflection coefficient 'Γ_g' which obtains at the output port or terminal 2. (Note that, although no provision has been made to stabilise the phase of b_3, this is of no consequence in a system with only a single source of microwave energy.) In addition to the directivity factor S_{32}/S_{31}, which is ideally zero, Γ_{gs} depends on S_{22}, which is, by definition, the reflection coefficient which obtains at port 2 when ports 1 and 3 have been terminated by matched (or reflection free) loads. The term S_{22} is also related to the 'main guide VSWR'[10] by the equation

$$Main\ Guide\ VSWR = \frac{1 + |S_{22}|}{1 - |S_{22}|} \tag{6.7}$$

Moreover, since $|S_{21}| < 1$, it is possible to infer an upper limit for Γ_{gs} from the directivity and main guide VSWR specifications. For example, given a 40 dB directivity, which is typical in a waveguide, one has $|S_{32}/S_{31}| = 0.01$, while a main guide VSWR of 1.05 yields $|S_{22}| = 0.025$. Assuming a worst case phase relationship in eqn. 6.6, one obtains a nominal 0.035 for the upper limit to $|\Gamma_{gs}|$. (As a practical matter, in the author's experience at least, the foregoing tends to give estimates of $|\Gamma_{gs}|$ which are considerably larger than anticipated (even after recognising their 'worst case' character) in comparison with the results obtained by 'direct measurement', as described in the paragraph which follows. The reason for this anomaly is not known.)

As an alternative to the above, it is also possible, and usually preferable, to determine Γ_{gs} by the following technique. As shown in Fig. 6.3, let the generator be connected to port 2, while port 1 is terminated by a load of variable reflection coefficient, which includes an adjustable transformer whose value is such that the signal b_3 (or power P_3) vanishes. Since eqn. 6.5 holds for all possible excitations, the reflection coefficient b_2/a_2, which is now obtained at terminal 2, is just Γ_{gs} and may be measured by any convenient method.

---

[10]For a definition of '*VSWR*' see Chapter 10

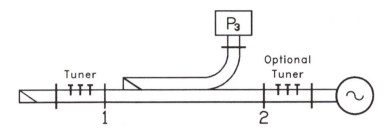

Fig. 6.3 Technique for measuring Γ_{gs}

Moreover, if desired, a two-port of adjustable parameters, or 'tuning transformer', may be added to port 2 to obtain the (usually desired) condition $\Gamma_{gs} = 0$.   This is an example of the use of a tuning transformer to provide a correction of the coupler imperfections, other examples will be included in the chapters which follow.   Once this has been accomplished, the termination on port 1, together with its adjustable transformer, may be discarded.   The foregoing will be recognised as the (conceptual) counterpart of the low-frequency technique of measuring source impedance after replacing the internal voltage source by a short circuit.

It is important to recognise in the above that the only role played by the transformer (or tuner) on port 1 was to establish the specified and uniquely defined excitation condition under which Γ_{gs} could be either measured or adjusted as required.   Once this is accomplished, however, the operation is, in principle, invariant to its further adjustment or removal. (It may, however, play a secondary role as described below.)   Although the insensitivity of Γ_{gs} to a tuner on port 1 (or port 3) could be proved in a formal manner, it will prove instructive to use an intuitive argument.   The physical basis may be recognised by postulating the permanent addition of a tuner at port 1 and of arbitrary adjustment.   According to the above, however, the observation or adjustment of Γ_{gs} calls first for establishing the excitation condition where $P_3 = 0$.   This in turn corresponds to a unique relationship between the forward and reverse wave amplitudes in the main guide which is usually most conveniently realised by an adjustable transformer on port 1.   If, as assumed above, a tuner of arbitrary and fixed characteristics is already included at port 1, and is by definition part of the three-port, it will usually be possible, and in any case necessary, for the one associated with the termination to also compensate for the presence of this added tuner such that the prescribed excitation conditions in the main body of the coupler are again realised. These, in turn, uniquely determine the observed reflection coefficient at port 2.

Because of its importance, an alternative point of view may also be useful. In particular, the lack of dependence on the generator impedance may be inferred directly from its absence in eqn. 6.5. From this the insensitivity of Γ_{gs}, to the addition of a tuner to the coupler at port 1, may be recognised as follows: Returning to Fig. 6.1, let a tuner be inserted at terminal 1. At this point one must decide whether the tuner is to be regarded as part of the generator or of the directional coupler. In either case, however, the *physical behaviour* of the combination cannot depend upon an arbitrarily chosen convention as to the dividing line between the 'generator' and 'coupler' sub-assemblies. Thus, if the operation is independent of the tuner when considered as part of the generator, the same *must be true* when it is considered as part of the coupler.

As a further detail, and this may be confirmed by inspection of eqn. 6.5, the value of Γ_{gs} is independent of Γ_d and thus to the addition of a tuner at port 3. In addition, however, this may also be recognised by recalling the initial step which called for an adjustment such that $P_3 = 0$. Presumably, in the absence of a signal at port 3, the observed impedance at port 2 cannot depend upon how port 3 has been terminated. This latter argument will prove useful in alternative formulations where the dependence on the detector impedances has been suppressed.

In practise, and as noted above, it is possible for a tuner at port 1 to play a useful secondary role which will now be explained. Returning to Fig. 6.2, it will be initially assumed that the different components, generator, modulator, coupler etc. are all 'ideal'. By hypothesis, there are no amplitude fluctuations in the generator and any reflected wave from the termination will be totally absorbed, either by the modulator or by the signal source. Thus P_3 will be constant and no action is required by the feedback loop. Next let a tuner, of substantial reflection, be inserted between the modulator and coupler. At this point, if a strongly reflecting termination is substituted for a matched one on port 2, there would be, except for the feedback action, a change in P_3. To be more specific, b_3 now includes a component which has been first reflected by the termination, then re-reflected by the tuner and finally added to the emerging wave from the modulator. It is to this composite signal that b_3 is proportional. The feedback action may thus be visualised as one which so adjusts the amplitude of the wave transmitted by the modulator that, when added to the 're-reflected' wave from the termination, the magnitude of their sum is constant. In an actual application, the feedback loop is generally called on to compensate for fluctuations in generator amplitude and, because of nonideal characteristics of the other components, for changes in the reflection coefficient of the termination on port 2. In practise, the demands placed on the feedback loop by the latter can be usually reduced, if not substantially eliminated, by means of a tuner at port 1 of the coupler. The desired tuner adjustment may be recognised by observing the magnitude of the corrective action taken by the feedback loop in response to a termination of highly variable reflection, usually a sliding short circuit, at port 2. The desired adjustment has been realised when no corrective action is observed or required in response to this stimulus. In this way, by keeping the dissipation in the control element nominally constant, it is ordinarily possible to reduce its magnitude and thus achieve an increase in the available stabilised power. Moreover, another potential problem is also minimised: The impedance presented to the generator by the modulator may be a function of its state of excitation or attenuation. To the extent that this can be kept constant, the potential for interaction with the source frequency, for example, is reduced.

Measurement and application of P_3

The signal b_3, or its counterpart P_3 can be measured in a variety of ways. In addition to the application described above, the question to be here considered is: 'In what way can these observations be used?' From the above it will be recognised that, in the absence of the feedback loop, fluctuations or changes in P_3 may be due *either* to amplitude instability in the signal source *or alternatively* to changes in the terminating impedance. Moreover, with the feedback circuit in use, no attempt is made differentiate between the two. The feedback simply keeps P_3 constant, irrespective of the source of a potential change. Taking a clue from this, it is convenient to postulate, in the absence of feedback, a mode of operation where all variations in P_3 are interpreted as changes in 'b_g' even though the actual source of some of them may be due to changes in load impedance. If this is done, the operation may be represented by a model characterised by a source reflection coefficient Γ_{gs} and where the 'b_g', although not constant, has been provided with a monitor so that changes in its magnitude can be observed. Although not quite as satisfactory as a stabilised source, this mode of operation may still be useful in certain measurement or calibration applications.

Multi-channel isolation

As an alternative to amplitude stabilisation, certain types of measuring systems use two or multi-channel configurations in order to obtain isolation from source instability and/or other benefits. Here it is generally required that the signal delivered to one channel is independent of changes in impedance or other loading effects in the second channel.   A common problem is that of achieving or determining that an adequate degree of isolation exists. Procedures for obtaining or recognising this condition will now be described.

Returning again to Fig. 6.1, but where the detector on port 3 has been removed, the desired isolation will be realised if the emerging wave amplitudes from ports 2 and 3 (b_2 and b_3) are constant. Ordinarily, however, it is *sufficient* that the *complex ratio* is constant. Thus

$$b_2/b_3 \ = \ K \tag{6.8}$$

where K is a complex constant of finite and nonzero magnitude.

This does not ensure that b_2, for example, will be independent of changes in the load impedance on arm 3. What it requires is that, if a change does occur, it will be accompanied by a simultaneous change in b_3 such that the net effect is that of a change in the amplitude and/or phase of the signal source and to which the operation is presumably insensitive.

Substitution of eqns. 6.3 and 6.4 in eqn. 6.8 now yields

$$(S_{21} - KS_{31})a_1 + (S_{22} - KS_{32})a_2 + (S_{23} - KS_{33})a_3 \ = \ 0 \tag{6.9}$$

Here, a_1, which represents the emergent wave from the generator, obviously does not vanish. Moreover, a_2 and a_3 may assume arbitrary and independent values. Thus the criterion for isolation requires that each of the coefficients for the a_i in eqn. 6.9 is zero.   First, with regard to the coefficient of a_1, neither of the terms S_{21} or S_{31} can vanish since this would mean no transmission between the respective ports and the generator. Thus, K is given by

$$K \ = \ S_{21}/S_{31} \tag{6.10}$$

After substituting this result in the coefficients for a_2 and a_3, one obtains

$$S_{22} - \frac{S_{21}S_{32}}{S_{31}} = 0 \tag{6.11}$$

and

$$S_{33} - \frac{S_{31}S_{23}}{S_{21}} = 0 \tag{6.12}$$

Comparison of eqn. 6.11 with eqn. 6.6 now indicates that the first of the conditions for isolation is that $\Gamma_{gs} = 0$. Assuming that the dividing network is a directional coupler, for example, this can usually be achieved by the addition of a tuner at port 2 as described above. Moreover, as was also noted, this condition is invariant to the addition and similar adjustment of a tuner on port 3, thus eqn. 6.12 may also be satisfied.

In principle, the extension of these results to three or more channels is straightforward, although as a practical matter certain 'internal' adjustments to the dividing network are also required. As an example, reference is made to Fig. 6.4, which includes a pair of

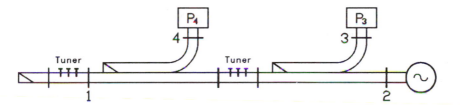

Fig. 6.4 Circuit for discussion of 'multi-channel' isolation

directional couplers in cascade and where a tuner, whose role will soon become apparent, has been inserted between them. As before, with excitation applied at port 2, one seeks an adjustment of the termination on port 1 such that the signals at *both* ports 3 and 4 simultaneously vanish. In general, however, it is not possible to realise both criteria by the adjustment of the termination alone. Instead, one adjusts the port 1 tuner for a null at port 4 and then adjusts the 'internal' tuner (between the two couplers) such that the signal at port 3 vanishes. At this point additional tuners may be added to ports 2, 3 and 4, to achieve the conditions given by eqn. 6.11 etc. In contrast to the earlier discussion about the lack of a continuing role for the tuner on port 1, however, the internal tuner adjustment must be maintained if the desired performance is to be realised. An application of these ideas will be made in Chapter 11.

Chapter 7

The measurement problem
at microwave frequencies

Although far from being all inclusive, it has been the author's objective in the preceding chapters to provide an introduction to the basic elements of microwave circuit theory. This model, in turn, has served to identify certain parameters, e.g. *power, attenuation, efficiency, reflection coefficient, scattering parameters* etc., whose determination is the objective of microwave metrology. Although further details will be added from time to time, it is believed that the foregoing chapters have provided an adequate motivation for the material to follow. This chapter will include a brief summary of some of the more important ideas and attempt to set the stage.

Measurements at microwave frequencies pose a unique set of problems for the metrologist. Generally speaking, and as emphasised in the earlier chapters, the component dimensions are comparable to the wavelength and retardation cannot be ignored. In the early art it was frequently asserted that electromagnetic field theory was the most appropriate tool for the description and evaluation of the projected and emerging microwave systems. It was not long, however, until certain similarities between the wave propagation in a uniconductor waveguide and that in the familiar TEM coaxial line were recognised. This quickly led to generalisations of 'voltage', 'current', etc. and to a 'microwave circuit theory', outlined in the preceding chapters, which in a formal sense is identical to that in use at the lower frequencies. Today, this microwave circuit theory plays a major role in the design and evaluation of microwave systems, while the field theory is used primarily in those applications where the greater detail which it provides is specifically required, as in the design of microwave components such as antennas or filters, for example. In spite of the similarities between microwave and low-frequency circuit theory, however, the subject of microwave measurements exists as a unique discipline. In order to better appreciate the reasons for this, some additional observations are useful.

The role of uniform waveguide

As outlined above, the components of a microwave system are usually interconnected via a transmission line or waveguide which, ideally, is of uniform cross-section and free of dissipation (lossless). Under these conditions, except for the values of two[11] complex constants, the fields are everywhere determined throughout the waveguide by its cross-sectional geometry and the parameters ϵ, μ of the enclosed media. These two complex constants may be the (generalised) voltage v and the current i at a given transverse plane in the waveguide, but more often are chosen to represent the (complex) amplitudes (b and a) of the wave propagation in the forward and reverse directions. Although there is a simple linear relationship between these two sets of parameters, there are important practical considerations which usually favour the use of b and a over v and i. The first of these relates to how these parameters change as a function of position along the waveguide. In contrast to v and i, in which both magnitude and phase vary with position, the magnitudes of the wave amplitudes (b and a) remain constant; it is only the phase which varies. Moreover, this phase variation is linearly related to the distance, which is not true for that of the generalised voltage or current (v and i).

Expression for power

Another consideration relates to the expression for the (net) power P. In the v, i description one has

$$P = \text{Re}(vi*) \qquad (7.1)$$

while in terms of b and a

$$P = |b|^2 - |a|^2 = |b|^2\left[1 - |\Gamma_l|^2\right] \qquad (7.2)$$

where $\Gamma_l = a/b$. In the former case, one requires the magnitudes of both v and i *plus* their phase difference, while for the latter, only the magnitudes are required. In addition, for reasons to be explained below, the measurement of power assumes a more basic and fundamental role. In particular, the question of 'how much' signal is present is far more likely to be answered by a power rather than 'voltage' measurement.

---

[11]Single mode excitation is assumed

Finally, the circuit parameters (impedance or admittance) associated with the v, i description are closely related to open or short-circuit conditions while the scattering parameters, associated with b and a, are defined in terms of 'matched' or reflectionless conditions. The latter usually correspond to system design objectives while operation under the former conditions may be untenable because of practical constraints. Thus, in many cases, the circuit (scattering) parameters associated with the b and a description are more easily measured.

Requirement for 'indirect observation'

This shift from v and i to b and a as the terminal variables, while characteristic of the microwave art, is not what causes microwave measurements to assume a different character. Of greater importance, in this context, is the more fundamental role assumed by the parameter *power*. Perhaps the greatest distinction, however, is that *it is generally impossible to make direct observations of the terminal variables!! (b, a)*. Not only are these ordinarily confined within a metallic boundary[12] but, of more fundamental concern, any attempt to introduce probes at the position of interest destroys the waveguide uniformity postulate, to which microwave circuit theory owes its existence. Ordinarily, one is prepared to go to substantial lengths to preserve this uniformity. This precludes the introduction of field sampling at the terminal planes of interest.[13] Measurements at microwave frequencies are thus usually in the context of a uniform transmission line and, while the associated complex wave amplitudes b and a play a major role in microwave circuit theory and analysis, in a practical sense their values are usually inferred only by indirect methods such as power measurement.

This, in turn, leads to other problems which are characteristic of the microwave art and which are perhaps best illustrated by specific examples. Referring to Fig. 7.1, one has a microwave system comprising a signal source, an intervening two-port and a load. Two terminal planes have been identified and labelled 1 and 2. The associated wave amplitudes are denoted by a_1, b_1, a_2, b_2 as shown. In

---

[12] An exception is the parallel wire transmission line, but this is of limited interest at microwave frequencies.

[13] This may be permissible where the ultimate in accuracy is not required. The trend, however, is away from such methods.

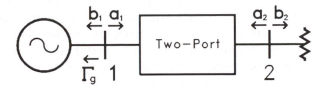

Fig. 7.1 Generator, two-port, and termination

many cases, absolute values for the phase of these terms are of no interest, but only phase differences. Thus a 'complete' specification of the fields at terminal 2, for example, involves only $|b_2|^2$ and the complex reflection coefficient a_2/b_2. The latter term is determined solely by the load properties. Quite generally, $|b_2|^2$ is determined by the level of the applied excitation, is sometimes referred to as an 'active' circuit parameter and is identified with power measurements. The reflection coefficient a_2/b_2, on the other hand, is a property of the microwave hardware, is sometimes called a 'passive' parameter and is identified with impedance measurements. If $a_2 = 0$, the termination is said to be 'matched' whereas if b_2 is independent of a_2, the source, as it appears at port 2, is said to be 'matched'. (As noted above, however, other meanings are also ascribed to the term 'matched'.)

Power measurement

Consider now the practical problem of measuring the power delivered to the termination or load. Since one cannot observe b_2 or a_2 directly, this calls first for the replacement of the load by a power meter which, ideally, reads $|b_2|^2$. If the source is 'matched', $|b_2|^2$ will not change when the load is reconnected. Moreover, if the reflection coefficient (Γ_l) is known for the load, the power may be determined by the use of eqn. 7.2. In the more general case, where the source is not matched, one must determine the parameters of its 'Thevenin equivalent circuit', (or $|b_g|^2$ and Γ_g) which requires three parameters. This, together with the complex impedance of the load, permits one to determine the power for an arbitrary set of source and load parameters. In the process, however, it has been necessary to determine one real and two complex parameters in order to obtain a single scalar parameter (power).

Another common problem is to determine the change in power which results in Fig. 7.1 if the two-port is removed. Provided that the generator (now referred to port 1) and load are both 'matched', the answer is simply given by the 'attenuation' of the two-port which is a single scalar parameter.

As will be explained in Chapter 11, when the 'matched' conditions do not obtain, the problem is far more difficult. The answer now involves the complex impedance of both generator and load, *plus* the three additional complex (scattering) parameters which describe the two-port (four if the device is nonreciprocal).

One method of coping with the inability to make direct observations of *b* and *a* is thus to evaluate the individual components in sufficient detail so that the performance of the composite system may be predicted by the appropriate circuit theory. In many cases, however, these additional details are of little or no further interest so that their measurement represents an undesirable, but usually unavoidable, burden on the metrologist.

Measurement strategies

In the evaluation of the system components, the arrangement of Fig. 7.2 will prove to be very useful. This circuit is typically based on one or more directional couplers as described in Chapter 6. For

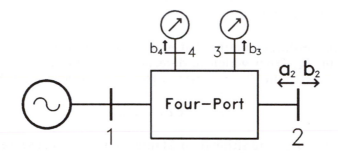

Fig. 7.2 A typical and useful measurement setup

example, the 'internal termination', which is found in the common three-port coupler, may be replaced by second detector. In any case, the object is to measure b_2 and a_2 which, in turn, determine the reflection coefficient and power delivered to the termination. If the item to be evaluated is a two-port, a second 'measurement network' may be added at its output and the two-port parameters determined from the input and output wave amplitudes under several different conditions of excitation. The key to all this is the measurement of b_2 and a_2. As already noted, these are usually inferred from observations of the fields (b_3 and b_4) which exist at positions remote from the measurement plane.

Referring to Fig. 7.2, as will be shown below, one has

$$b_3 = Aa_2 + Bb_2 \qquad (7.3)$$

where b_2 and a_2 have already been defined, A and B are complex constants and b_3 is a measure of the complex field amplitude at position 3 in the measurement circuit. Ordinarily b_3, in common with b_2, is interpreted as an emerging wave amplitude. In this case eqn. 7.3 may be obtained as a solution of the scattering equations.   In particular, A and B may be expressed in terms of the scattering coefficients and the reflection coefficients of the detectors which terminate arms 3 and 4 of the 'measurement network'. (This result may also be compared with eqn. 6.5.) As will be demonstrated in the next chapter, however, depending upon how A and B have been defined, b_3 may actually represent any linear function of the complex electric and magnetic field amplitudes at the indicated position. Moreover if, as assumed above, the coupling between the signal source and the measurement network in Fig. 7.2 is via a single 'mode', then A and B are determined entirely by the properties of the measurement network.

At this point, assuming that A and B are known, one has a single observation, b_3, from which to determine the unknowns, a_2 and b_2. Quite obviously, this is impossible. However, at arm 4 one also has

$$b_4 = Ca_2 + Db_2 \qquad (7.4)$$

where C and D are additional constants. Given the observed values of b_3 and b_4 and assuming that $A \cdots D$ are known, one can, in theory, easily solve eqns. 7.3 and 7.4 for b_2 and a_2. Generally speaking, the role of the metrologist includes not only the observations of b_3 and b_4 but the determination of $A \cdots D$ as well. This is usually, by far, the most difficult part of the procedure.

Although conceptually straightforward, prior to the advent of automation, this approach found little practical application for at least two reasons. Not only is a substantial amount of computation involved but, because b_3 and b_4 are required on a complex basis, a fairly sophisticated detection system is required. Instead, much of the early and continuing art is built on the much simpler detection systems (e.g., diodes or bolometers) which yield magnitude information only.

If diodes or bolometers are used as the detectors, eqns. 7.3 and 7.4 become

$$P_3 = |b_3|^2 = |Aa_2 + Bb_2|^2 \qquad (7.5)$$

$$P_4 = |b_4|^2 = |Ca_2 + Db_2|^2 \qquad (7.6)$$

where P_3, P_4 are measured powers and the proportionality factors have been absorbed by $A \cdots D$. At this point, unfortunately, the system is no longer 'determinate', since one has only two scalar observations, P_3 and P_4, from which to determine the *complex* a_2 and b_2. At the very least, if $|b_2|^2$ and the *complex* ratio a_2/b_2 are taken as the desired measurands, one requires a minimum of three experimental observations. At the same time, there is no law of nature which limits the number of detectors to two! In a later chapter the 'six-port' technique will be described, which is based on a six-port measurement network and the use of *four* power detectors. The more elementary techniques, which were in general use prior to the advent of these methods, may still be described by eqns. 7.5 and 7.6, in which, as developed in the chapters to follow, the $A \cdots D$ satisfy certain constraints. For example, if $|A| = |B|$ and the phase between them is adjustable, one has a 'slotted line'. On the other hand, if $B = C = 0$, this provides the basis for a 'reflectometer'.

Comparison with lower frequency art

In summary, and in contrast with the practise at lower frequencies, the microwave metrology environment is one in which, for all practical purposes, the voltmeters and ammeters have been discarded in favour of energy-sensing devices. Since these usually absorb the power they measure, their loading effects must generally be taken into account by the metrologist. As a rule, it is not possible to make 'direct' observations at the measurement port. Instead, as indicated by Fig. 7.2 and described by eqns. 7.3 $\cdots$ 7.6, the response of detectors which are 'remote' from there must be processed by mathematical methods to yield the information of interest. The role of the metrologist generally includes the determination of the parameters, e.g. $A \cdots D$, which describe the measuring instrument. In practise, and as already noted, this may be the most difficult part of the procedure. Moreover, because these instruments are not generally noted for their stability, a substantial premium is placed on methods for doing this quickly and accurately.

Sliding terminations

The microwave metrologist does, however, have at his disposal a tool which is without a counterpart in the low-frequency art: the *sliding termination* or *sliding load*. As shown in Fig. 7.3, these

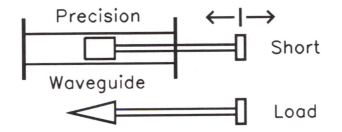

Fig. 7.3 Sliding terminations play an important role in microwave metrology.

devices consist of a section of transmission line and include a moving plunger whose dimensions complement those of the line such that longitudinal motion is possible, but with a minimum of 'wobble'. Ideally, these provide a reflection of constant magnitude and variable phase, and where the phase change is proportional to the linear displacement. Although weakly reflecting sliding terminations are more readily available, substantial use has also been made of strongly reflecting ones or 'sliding shorts', particularly in metrology laboratories. The application of these devices will be described in the chapters to follow.

In summary, eqns. 7.3 ⋯ 7.6, including the specialised cases as noted, provide the theoretical basis for much of the art of microwave metrology. In the following two chapters, their derivation will be given in greater detail and certain properties of the linear fractional or bi-linear transform, which is a direct consequence of them, will be presented. The succeeding chapters will, to a large measure, be devoted to a description of the methods by which these ideas are exploited.

Chapter 8

The fundamental equation of microwave metrology

In recognition of its all pervading nature and widespread application, eqn. 7.3, which may also be written

$$b_i = A_i a_2 + B_i b_2 \qquad (8.1)$$

will be given the title '**The fundamental equation of microwave metrology**'. It is the primary objective of this chapter to provide its derivation under a fairly general set of initial conditions. This task, in turn, will become a vehicle for developing certain analytical techniques which find repeated application in microwave metrology. In the process, eqn. 8.1 will actually be obtained via several different methods, each of which provides a different perspective.

To begin, reference is made to Fig. 8.1 where an n-port junction is assumed. By convention, port 1 is connected to the generator. The initial objective, which is the subject of this chapter, is to obtain the

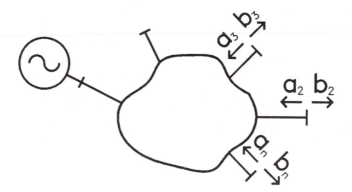

Fig. 8.1 Generator and 'n'-port junction

response of the measuring instrument in terms of the emergent and incident wave amplitudes, b_2 and a_2, which characterise the excitation state at port 2 and which is also known as the 'test' or 'reference' port. In later chapters, the problem of 'inverting', or solving the resultant equations for the excitation conditions, or parameters of interest, will be considered. As noted in Chapter 7, it is usually sufficient to determine $|b_2|^2$ and the ratio a_2/b_2, or reflection coefficient Γ_l which exist at the test port. Indeed, for many metrology applications, the determination of Γ_l alone is sufficient and in general b_2 may be eliminated from the problem by forming the ratio between the responses of two detectors.

The scattering equations for Fig. 8.1 may be written

$$b_1 = S_{11}a_1 + S_{12}a_2 + \cdots + S_{1n}a_n$$

$$\vdots \qquad\qquad\qquad\qquad (8.2)$$

$$b_n = S_{n1}a_1 + S_{n2}a_2 + \cdots + S_{nn}a_n$$

In addition, and by hypothesis, each of the remaining ports (other than 1 or 2) are terminated by detectors which yield either b_i, P_i (or $|b_i|^2$) or possibly a *ratio* among the b_i, e.g. b_3/b_4.[14] In addition, these detectors are characterised by a reflection coefficient Γ_{di} such that

$$a_i = b_i\Gamma_{di}, \qquad i = 3 \cdots n. \qquad (8.3)$$

Although the solution of eqns. 8.2 and 8.3 for eqn. 8.1 may be achieved by elementary algebraic methods, this approach becomes tedious and unwieldy if two or more detectors are involved. Moreover, the resultant expressions, which are functions of the scattering parameters and detector reflection coefficients, are so complicated that the more basic relationships tend to be obscured. In keeping with the philosophy espoused in the introductory chapter, the objective is that of formulating the problem in such a way as to answer the questions of interest – no more no less. In this case, the result of primary interest may be obtained with little effort by the application of certain well known results from linear algebra.

At each of the terminal surfaces, the excitation conditions are specified by the b_i and a_i, where $i = 1 \cdots n$. Although these represent

---

[14]Strictly speaking, proportionality factors should be included, but these may be 'absorbed' by the A_i and B_i.

$2n$ variables, in general the constraints imposed by the associated network are such that one half of them are determined by the remaining half. The scattering equations, for example, provide the b_i as functions of the a_i. The $2n$ variables are thus connected by n equations of constraint. In addition, however, the $(n-2)$ detectors provide an additional $(n-2)$ constraints of the form $a_i = b_i \Gamma_{di}$ (eqn. 8.3). Thus, in all, one has $(2n-2)$ constraints on the $2n$ variables. From linear algebra (and apart from possible singularities!!) it is known that $(2n-2)$ of these variables can be solved as a linear function of the remaining two. This leads directly to eqn. 8.1 where a_2 and b_2 are taken as the independent variables.

An alternative derivation

Because it is based on a scattering equation formulation, the foregoing implicitly assumes, however, that the different ports are provided with uniform waveguides such that the b_i and a_i are 'well defined'. As will next be demonstrated, the only place where this is really required is at the measurement port.

In Fig. 8.2 let there be an *enforced* current $a_1 \mathbf{J}$ in one portion of the structure where $\mathbf{J}$ is a three-dimensional vector function giving the spatial distribution and where a_1 is a scalar multiplier. The

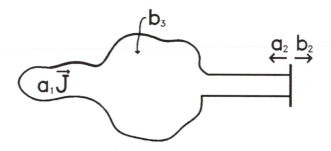

Fig. 8.2 'Circuit' for alternative derivation

current $a_1 \mathbf{J}$ is the source of an electromagnetic field throughout the structure and is, by hypothesis, independent of the termination on the uniform section.

Let the uniform section (arm 2) be first terminated by a matched load. Then the field in arm 2 is completely specified by the term b_2 which is a measure of the electric field amplitude associated with the outgoing travelling wave. In another part of the structure let b_3 represent some *linear* function of the complex electric and magnetic field amplitudes. If the system is *linear* it is possible to write

$$b_2 = ha_1 \tag{8.4}$$

$$b_3 = ka_1 \tag{8.5}$$

where h and k are constants.

Next, let an incoming wave, the amplitude of which will be denoted a_2, be launched in arm 2.[15] In general this wave will be partially reflected, giving rise to an additional component in b_2. A similar observation holds for b_3. By superposition, b_2 and b_3 now become

$$b_2 = ha_1 + ma_2 \tag{8.6}$$

$$b_3 = ka_1 + na_2 \tag{8.7}$$

where m and n are two additional constants. Elimination of a_1 between eqns. 8.6 and 8.7 yields

$$b_3 = A_3 a_2 + B_3 b_2 \tag{8.8}$$

where

$$A_3 = n - km/h \tag{8.9}$$

and

$$B_3 = k/h \tag{8.10}$$

In general, this argument may be repeated for any position (i) in the structure. This leads to eqn. 8.1.

Discussion

In addition to demonstrating that eqn. 8.1 obtains in a fairly general context, both of these techniques lead directly to the system response as a function of the parameters to be measured. In particular, the dependence on the network and detector parameters has been largely suppressed. Thus, if the problem includes that of assessing the effect of an impedance change in one of the detectors, for example, the given formulation or model is no longer adequate to answer the questions of interest. Moreover, the solution provides only

---

[15]This may be achieved *either* by a source on arm 2, *or* by a partial reflection of the wave b_2. The argument is the same in either case.

a limited amount of insight into what physically determines the A_i and B_i. They are, of course, functions of the network, but what about the generator, where the boundary between it and the network is also arbitrary?

To answer this question, it is useful to begin with Fig. 8.3. By hypothesis one has a source (left) and passive network (right) which are interconnected via a single mode waveguide. Within this guide the

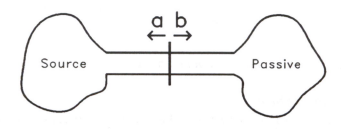

Fig. 8.3 Source, <u>single mode</u> waveguide, and passive termination

fields are completely determined by the wave amplitudes, a and b, therefore the same is evidently true for those in the passive network. However, one also has $a = b\Gamma_l$, which is independent of the generator, so that the fields are completely determined by b. In particular, the *ratio* of the field amplitude at one position in the passive network to that at another is independent of b and thus of the generator characteristics. This assumes, however, that the interaction between source and network is limited to a single *mode*. If more than one mode is involved, it is necessary to represent the interaction as indicated in Fig. 8.4 which includes multiple connections between source and network. In particular, the field which now exists in the passive termination may be represented as a linear combination of two separate fields. The first is that which obtains for a postulated excitation where $b_1 = 0$, $b_2 \neq 0$. The second corresponds to $b_1 \neq 0$, $b_2 = 0$. If the three-port network to the left of the terminal planes is part of the generator, the field distribution in the network on the right will certainly be dependent on the 'generator' characteristics. Moreover, if additional 'ports' are added so that it becomes the 'measurement network', the A_i and B_i will certainly depend on the generator. Conversely if, as suggested in Fig. 8.3, the generator is 'separated' from the remainder of the network by something which approximates a single mode structure, it is generally safe to assume that A_i and B_i will be independent of the generator properties.

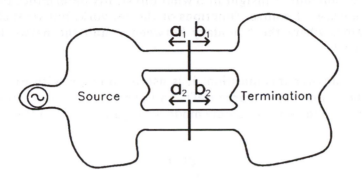

Fig. 8.4 Multi-mode interaction between source and termination

A third derivation

As noted above, however, the foregoing provides little, if any, insight into what physically determines the A_i and B_i. In order to address these concerns, another treatment, which provides a

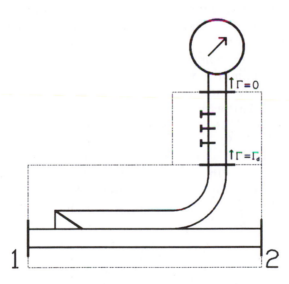

Fig. 8.5 Directional coupler and sidearm detector

substantial amount of additional insight, is also of interest. Referring to Fig. 8.5, a three-port network, in this case a directional coupler and detector are shown. Here one can *model* the detector as a series combination of a matched termination, preceded by a tuning transformer whose reflection duplicates that of the detector. Now, referring again to Fig. 8.5, once the device has been assembled, the choice of what constitutes the 'detector' as opposed to the 'directional coupler' is arbitrary. Thus if one chooses to make the (fictitious) tuning transformer part of the coupler rather than detector, the scattering equations become

$$b_2 = S_{21}a_1 + S_{22}a_2 \qquad (8.11)$$

$$b_3 = S_{31}a_1 + S_{32}a_2 \qquad (8.12)$$

since, by hypothesis, a_3 is now zero.

Eliminating a_1 between eqns. 8.11 and 8.12 yields

$$b_3 = \frac{S_{31}}{S_{21}} b_2 - \frac{S_{31}}{S_{21}} \left(S_{22} - \frac{S_{32}S_{21}}{S_{31}} \right) a_2 \qquad (8.13)$$

which may be compared with eqns. 8.1 and 6.5. Moreover, the same technique may be applied to Fig. 8.1 which leads to a similar result in a multi-detector environment.

It is important to take a hard and careful look at what has been done here. In general, one is only interested in the performance of the composite system. It does include several components, but after assembly one can choose the boundary between these to suit one's computational convenience. Quite apart from the results of the prior analysis, the outlined procedure leads to a correct and completely general description of the operation of the *composite* system, but under the simplifying assumption of matched detectors. On the other hand, it is equally important to recognise that if the problem were that of predicting the system performance, given the characteristics of the individual components, this technique would not work. In the microwave art, however, the prevailing practise generally calls for first assembling the system and then asking for its composite properties, rather than the converse. As a corollary to this, one may observe that, although the scattering equation formulation plays an important role in the description of microwave systems, the scattering coefficients, themselves, are seldom the measurement objective. More often one is interested in some function of them, frequently including the detector parameters as well. The equivalent source reflection coefficient for a stabilised generator, which was introduced in the prior chapter, provides a good example.

Thus, by adopting the foregoing convention, one can substantially simplify the algebraic effort required to obtain the solution to what may be a fairly comprehensive problem. At the same time, it must be recognised that this puts the terminal plane between the coupler and detector in a 'physically inaccessible' position and, if the problem includes that of evaluating the effect of replacing one detector by another, for example, a more general analysis is required. It must be emphasised that the 'scattering parameters', which appear in eqns. 8.11 ⋯ 8.13 are not those of the coupler alone, but also include the detector characteristics as well. In practise, however, these latter effects are frequently small and eqn. 8.13 will often provide a fairly accurate indication of the response if the 'scattering coefficients' are assumed to be those of the coupler alone.

Chapter 9

Properties of
The linear fractional transform

As noted in the previous chapter, in many cases the measurand of interest is just the ratio a_2/b_2 or Γ_l, which exists at the test port. Thus, after taking the ratio of the equations which result from letting $i = 3$ and $i = 4$ in eqn. 8.1, one obtains

$$\frac{b_3}{b_4} = w = \frac{a\Gamma_l + b}{c\Gamma_l + 1} \tag{9.1}$$

where $a = A_3/B_4$, $b = B_3/B_4$, $c = A_4/B_4$, $B_4 \neq 0$ and $\Gamma_l = a_2/b_2$. Equation 9.1 gives the relationship between w or b_3/b_4, which by hypothesis is 'observable', and the unknown parameter Γ_l which is here assumed to be the measurement objective. As noted in an earlier chapter, this functional relationship, which is the *ratio* between two *linear* functions of Γ_l, carries the name 'linear fractional' or 'bi-linear' transform. In view of its importance and all pervading presence in microwave metrology, some of its general properties will now be described. It should first be recognised that we are dealing with a functional relationship between two *complex* numbers, w and Γ_l, each of which requires a plane for its representation. For this reason it is not possible to 'plot' w vs Γ_l in the usual sense of the word. (To do so would require a four-dimensional space!) Instead, one uses the technique of 'mapping' or describing how a selected continuum of points or values of Γ_l, usually representing a two-dimensional figure in the 'Γ_l-plane', will appear when 'transformed' or 'mapped' into the 'w-plane' after imposing the functional relationship eqn. 9.1. In the general case, this can be a fairly time consuming and comprehensive procedure, but some examples may help.

Perhaps the simplest case is for $c = 0$, $a = 1$. Then, $w = \Gamma_l + b$ and the mapping requires merely a shift in origin between the two planes. Next, with $b = c = 0$, one has $w = a\Gamma_l$. If a is real and positive, the transform is merely a contraction or expansion of the co-ordinate axes by the factor a. In the general case, where $a = |a|e^{j\theta}$, an axis rotation of θ is also involved. Finally, let $a = 0$, $b = c$ and let $b \to \infty$ (or $B_4 = 0$ in eqn. 8.1). Then $w = 1/\Gamma_l$. In this case, if $|\Gamma_l| < 1$,

$|w| > 1$ since $w\Gamma_I = 1$. Thus, the entire region *inside* the unit circle, in the Γ_I plane, is mapped *outside* the unit circle in the w plane and conversely. In addition, the argument of w will be the negative of that for Γ_I. This operation is sometimes described as an 'inversion in the unit circle followed by a reflection in the real axis'. It can be shown that the general linear fractional transform is a combination of the elementary operations just described.

Circle mapping

One of the better known properties of the linear fractional transform is that 'circles are mapped into circles, with straight lines as limiting cases', and this is certainly the feature of greatest interest in the present context. Although this can be proved on the basis of the foregoing, a simpler derivation is as follows:

Let $\Gamma_I = re^{j\theta}$, where r is real and positive. Then, if r is held constant and θ permitted to vary, the locus of Γ_I is a circle of radius r and centred at the origin of the complex 'Γ_I-plane'. Starting with

$$w = \frac{are^{j\theta} + b}{cre^{j\theta} + 1} \tag{9.2}$$

after adding and subtracting the first term on the right in the expression which follows, one has

$$w = \frac{b - ac^*r^2}{1 - |c|^2 r^2} + \frac{(a - bc)re^{j\theta}}{1 - |c|^2 r^2} \left[\frac{1 + c^*re^{-j\theta}}{1 + cre^{j\theta}} \right] \tag{9.3}$$

Examination of this result reveals that the first term, on the right, is independent of θ and thus a constant since r is constant and only θ is permitted to vary. With respect to the second this is the product of two factors. The magnitude of the first is constant and its argument is given by $\theta + \psi$, where $\psi = \mathrm{Arg}(a - bc)$. The second factor is the ratio between two functions of θ, but since these are complex conjugates of each other, the magnitude of this factor is unity. Thus, the locus of w is a circle of radius R and centre R_c where

$$R = \frac{|a - bc|r}{1 - |c|^2 r^2} \tag{9.4}$$

and

$$R_c = \frac{b - ac*r^2}{1 - |c|^2 r^2} \qquad (9.5)$$

With regard to the expression $1 + cre^{j\theta}$, which occurs in eqn. 9.3, it will be evident from an inspection of Fig. 9.1 that its argument ϕ will be a function of θ which varies between $\pm\sin^{-1}(r|c|)$ where

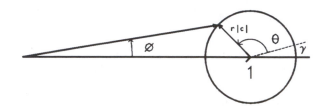

Fig. 9.1 Showing the relationship between ϕ and θ

$r|c| < 1$. (Note that the reference position γ, from which θ is measured, is equal to $Arg(c)$). In a practical application, θ will be a function of the longitudinal position of the sliding termination, the argument of its reflection coefficient etc. To be more specific, one has

$$\phi = \tan^{-1}\left(\frac{r|c|\sin\beta}{1 + r|c|\cos\beta}\right) \qquad (9.6)$$

where $\beta = \theta + \gamma$. The argument of the second term in eqn. 9.3 is thus given by

$$Arg\left\{\frac{(a - bc)re^{j\theta}}{1 - |c|^2 r^2} \cdot \left[\frac{1 + c*re^{-j\theta}}{1 + cre^{j\theta}}\right]\right\} = \theta + \psi - 2\phi \qquad (9.7)$$

As indicated by Fig. 9.2, the dependence of ϕ on θ is in the form of a 'distorted' sine wave. Thus, the total argument of the second term alternately leads and lags that of $\theta + \psi$ alone.

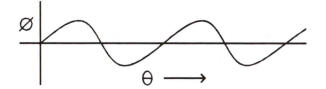

Fig. 9.2 The dependence of ϕ on θ is that of a 'distorted' sine wave.

Additional results

As noted, the foregoing results are those of primary interest. However for completeness, some additional features will also be included.   If θ is constant, while r is permitted to vary, the counterpart of eqn. 9.3 may be written

$$w = \frac{ae^{j\theta} - bc^*e^{-j\theta}}{ce^{j\theta} - c^*e^{-j\theta}} - \frac{(a - bc)e^{j\theta}}{ce^{j\theta} - c^*e^{-j\theta}} \cdot \left[\frac{1 + c^*re^{-j\theta}}{1 + cre^{j\theta}} \right] \qquad (9.8)$$

with a similar interpretation to that already given.   Here, the first term on the right is independent of r and thus a constant if only r varies.   The first factor in the second term is also independent of r, while the magnitude of the second is unity.   As before, the locus is a circle of radius R_θ and centre $R_{c\theta}$, where

$$R_\theta = \frac{|a - bc|}{|ce^{j\theta} - c^*e^{-j\theta}|} \qquad (9.9)$$

$$R_{c\theta} = \frac{ae^{j\theta} - bc^*e^{-j\theta}}{ce^{j\theta} - c^*e^{-j\theta}} \qquad (9.10)$$

In order to demonstrate that the locus does include a full circle it is convenient to define

$$F(r) = 1 + cre^{j\theta} \qquad (9.11)$$

and where, as already noted, $\phi = \mathrm{Arg}[F(r)]$.   The argument of the second factor in the second term is thus -2ϕ.   Referring to Fig. 9.3,

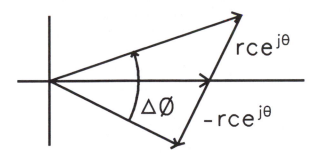

Fig. 9.3 The total excursion in ϕ is $180°$

it is evident that the total excursion in ϕ will be 180° (and thus 360° in 2ϕ) provided that r is permitted to take on *negative* as well as positive values. (Note that, if r is restricted to positive values, this defines a 'half-line' in the Γ_l plane, or, alternatively, that it is now necessary to include both θ and $\theta + 180°$ in order to get a complete circle.)

Returning to eqn. 9.5, if $r = 0$, then $R_c = b$. If $r \to \infty$, $R_c = a/c$. These two points play a major role in the transform geometry. Let $R_k = (b + a/c)/2$ represent their midpoint. Then the (vector) difference in going from R_k to R_c is given by

$$R_c - R_k = -\frac{(a - bc)}{2c} \left[\frac{1 + |c|^2 r^2}{1 - |c|^2 r^2} \right] \tag{9.12}$$

The argument of this expression is independent of r (and thus a constant). This indicates that the circle centers R_c lie on the straight line which connects b and a/c.

To continue, the ratio of the circle radius R to the distance between R_k and R_c $\left[R/|R_c - R_k| \right]$ is given by

$$\frac{R}{|R_c - R_k|} = \frac{2r|c|}{1 + |c|^2 r^2} = 1 - \frac{(1 - |c|r)^2}{1 + |c|^2 r^2} \tag{9.13}$$

which is always less than but approaches unity if $r|c| \to 1$. As a consequence, the point R_k is not enclosed by any of this family of circles. On the other hand, it is a simple exercise to demonstrate that $|b - R_c|/R = |cr|$. Thus, the point b is enclosed by those circles for which $r < |c|^{-1}$. In a similar way one can show that $|a/c - R_c|/R = |cr|^{-1}$. Thus the point a/c is enclosed by those circles for which $r > |c|^{-1}$.

It will next be demonstrated that the circles associated with the fixed or constant values of θ go through the points b and a/c. After subtracting b and a/c in turn from $R_{c\theta}$ one has

$$R_{c\theta} - b = \frac{(a - bc)e^{j\theta}}{ce^{j\theta} - c^*e^{-j\theta}} \qquad (9.14)$$

$$R_{c\theta} - (a/c) = \frac{c^*(a - bc)e^{-j\theta}}{c\left[ce^{j\theta} - c^*e^{-j\theta}\right]} \qquad (9.15)$$

Comparison of these results with the value of R_θ given by eqn. 9.9 indicates that the magnitudes are equal. Therefore, the distances from b and a/c are both equal to the circle radius (and thus to each other) which proves the given statement. As a further consequence, one concludes that the circle centers $R_{c\theta}$ are all found on the perpendicular bisector of the line which connects b and a/c.

The foregoing is summarised in Fig. 9.4. Those values of r for which $|c|r < 1$ are mapped into the half-plane which contains b and whose boundary is the 'perpendicular' which was identified above.

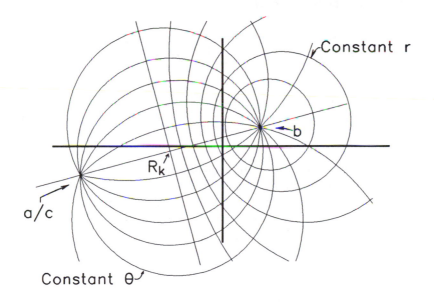

Fig. 9.4 Summary for circle mapping discussion

Those values of r for which $|c|r > 1$ are mapped into the other half. As a practical matter it is usually desirable to avoid the discontinuity which results if $|c|r \to 1$. In practise, this is ordinarily no problem since the usual design objectives call for $c = 0$ and r, which represents Γ_l, is at most of unit magnitude. Thus the region of practical interest is usually only that in the immediate the neighborhood of b and these more general features are frequently of limited practical consequence. There are applications, however, where this added description is of value.

Although the foregoing has focused on the locus of a circle centred at the origin of the Γ_l plane, it is a simple exercise to show that the same general results are true for any circle. In particular, let $\Gamma_l = re^{j\theta} + r_c$. Substituting this in eqn. 9.1 yields

$$w = \frac{are^{j\theta} + b + ar_c}{cre^{j\theta} + 1 + cr_c} \tag{9.16}$$

which is easily shown to be of the same functional form as eqn. 9.2. Finally the equation for the limiting case of a straight line may be written $\Gamma_l = h + kx$ where h and k are *complex* constants, while x is a *real* variable. Again, it is easily shown that this reduces to the case of a variable r and constant θ.

Chapter 10

The slotted line technique of impedance measurement

Even though it has been largely replaced by other methods, no text on microwave metrology would be complete without at least a brief treatment of the 'slotted line'.

Conceptually this is perhaps the most simple and certainly one of, if not the oldest method of measuring microwave impedance. The basic form of the measuring instrument includes a section of the appropriate transmission line, but in which a longitudinal 'slot' has been cut. This permits the fields to be sampled by means of a suitable probe in conjunction with a transport mechanism such that (hopefully!) the coupling is constant with respect to longitudinal position. Although it will be immediately recognised that the 'uniformity' criterion has been violated, it may be that the resultant field perturbations can be tolerated if the ultimate in accuracy is not required. In any case, the perturbation due to the slot can be minimised by keeping it narrow and parallel to the current paths. In the case of the rectangular waveguide, this requires that the slot is located in the centre of the broad wall. (On the narrow wall the current paths are transverse to the axis.) The other source of perturbation is due to the probe itself. In general this can be minimised by the use of a sensitive detection system and by keeping the depth of penetration, or coupling, to a minimum. The probe output is usually associated with a detector which responds to amplitude only or power P_3 such that

$$P_3 = \left| Aa_2 + Bb_2 \right|^2 \tag{10.1}$$

where A and B are as yet to be determined functions of probe position.

In the most common arrangement, the probe may be regarded as a (very) short 'receiving antenna', perpendicular to the waveguide axis and of adjustable length or penetration. Its (nondirectional) response

73

is primarily to the total electric field amplitude which exists at the probe position.

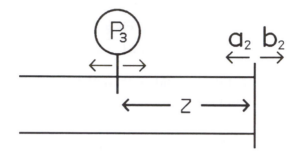

Fig. 10.1 Illustration of slotted line parameters

Referring to eqn. 3.2 and Fig. 10.1, P_3 will be given by

$$P_3 = K \,|\, b_2 e^{-j\beta z} + a_2 e^{j\beta z} \,|^{\,2}$$ (10.2)

where K is a proportionality constant and z is measured from the terminal plane. (Note that in Fig. 10.1 the actual values for z are negative.) Equation 10.2 may also be written

$$P_3 = K \,|\, b_2 + a_2 e^{2j\beta z} \,|^{\,2}$$ (10.3)

Comparison of this result with eqn. 10.1 indicates that $|A| = |B|$. Their phase (difference) and thus the probe response P_3 is a function of probe position. The maximum and minimum values of P_3 are given by

$$P_{3max} = K \Big[|b_2| + |a_2| \Big]^{2}$$ (10.4)

and

$$P_{3min} = K \Big[|b_2| - |a_2| \Big]^{2}$$ (10.5)

The square root of the ratio P_{3max}/P_{3min} is by definition the 'voltage standing wave ratio' (VSWR or sometimes just SWR). Thus, from eqns. 10.4 and 10.5, one has[16]

---

[16]In an earlier era, the reciprocal of this definition was also in use, but this has been discarded.

$$VSWR = \frac{|b_2| + |a_2|}{|b_2| - |a_2|} = \frac{1 + |\Gamma_l|}{1 - |\Gamma_l|} \qquad (10.6)$$

and

$$|\Gamma_l| = \frac{VSWR - 1}{VSWR + 1} \qquad (10.7)$$

In particular, the factor 'K', and with it the dependence on source level, has been eliminated as a consequence of forming the *ratio* between two power readings.

It may be further noted that, in taking the square root, it has been implicitly assumed that $|a| < |b|$. In general, as might be expected from symmetry, it is not possible, by this method, to determine the direction of energy flow (or to which end of the line the generator is connected). To do so requires a 'directive' coupling device, e.g. a *directional coupler*.

In addition, although a less common application, the slotted line may be used to obtain the argument of Γ_l. To demonstrate this it is convenient to write eqn. 10.3 in the form

$$P_3 = K|b_2|^2 |1 + \Gamma_l e^{j4\pi z/\lambda}|^2 \qquad (10.8)$$

where Γ_l has been substituted for a_2/b_2 and β has been replaced by $2\pi/\lambda$. At the probe minimum, $z = z_{min}$ and one has

$$\text{Arg}(\Gamma_l) + 4\pi z_{min}/\lambda = \pi \pm 2n\pi \qquad (10.9)$$

By inspection, the distance between successive minima is $\lambda/2$, since when added to z, this quantity increases the argument by 2π. Moreover, eqn. 10.9 (implicitly) provides $\text{Arg}(\Gamma_l)$ in terms of z_{min}, where z is measured from the output port. As a rule, however, it is more convenient to establish a reference position within the slot itself. This may be done by first observing the value of $z_{min} = z_s$ with the reference port terminated by a short and for which $\text{Arg}(\Gamma_s) = \pi$. If $z_{min} = z_l$ for Γ_l, one has from eqn. 10.9

$$\text{Arg}(\Gamma_l) + 4\pi z_l/\lambda = \pi + 4\pi z_s/\lambda \qquad (10.10)$$

so that

$$\text{Arg}(\Gamma_l) = \pi\left[1 - 4(z_l - z_s)/\lambda\right] \qquad (10.11)$$

As before, the negative z direction is towards the generator and conversely.

As observed in the introduction, this technique played a major role in the earlier art. However, with increasing accuracy demands and the advent of automation, it has been largely replaced by other methods, although quasi-automated, broadband versions of the method may still be in use. Perhaps its major legacy to the current art is in the terminology which it has fostered. For example, one still finds adapters and attenuators specified in terms of the '*VSWR*' (in contrast to the possibly more logical value of $|S_{11}|$) which exists at one end when the other is terminated by an impedance match.

From the viewpoint of careful metrology, its sources of error include

(1) Impedance discontinuity introduced by the slot

(2) Inability to maintain uniform probe coupling to the fields due to imperfections in the (nonideal) transport mechanism

(3) Violation of the 'uniformity requirement', and distortion of field patterns due to probe 'loading'.

Of these problems, the last is possibly the most serious in that its magnitude is strongly dependent on the probe coupling and this does not easily lend itself to a quantitative assessment. In addition, an analysis in terms of voltage and current concepts indicates that the 'source impedance', as seen by the probe, is a function of probe position. Moreover, the loading effects are far more serious at the position of maximum response than at the minimum. In principle, one could account for some of these problems by a more general model which might be in terms of a moveable shunt admittance across the line. Presumably, this would then account for the probe reflection, probe-generator interaction etc. Unfortunately, however, the success of this scheme is contingent on a known value for the 'shunt admittance', or probe coupling, and this is not easily determined. Moreover it is a function of probe penetration. As noted above, this technique has been largely replaced by those described in the chapters which follow.

Chapter 11

Attenuation measurement

As briefly outlined in Chapter 5, a common two-port application is that of providing a reduction in power subsequent to its insertion in a given system. Provided that the system is *matched* at the point of insertion (i.e. $\Gamma_g = \Gamma_l = 0$) the power reduction is given by the *attenuation* (or $-20\log_{10}|S_{21}|$). When the stated criteria are not realised, the more general term is *insertion loss* which is, by definition,

$$\textit{Insertion loss (dB)} \; = \; -10\log_{10}(P_f/P_i) \tag{11.1}$$

where P_i and P_f are the initial and final powers respectively. (Note that in contrast to eqn. 5.10 the stipulation for a *matched* insertion point has been eliminated.)

From the historical perspective, no other aspect of microwave metrology has been characterised by a greater proliferation of techniques than attenuation. As was the case with the slotted line, this text would certainly be incomplete without a chapter on this subject. On the other hand, a comprehensive treatment calls for a *book*[17] rather than a chapter. To a substantial degree, moreover, this task has been taken over by the automated network analyser as described in the chapters to follow. For these reasons, this chapter will only provide a brief review of the prior art and identify certain basic features of the measurement task.

Power ratio methods

Perhaps the simplest technique (at least conceptually!) is that shown in Fig. 11.1. Here one has a measurement system which is comprised of a generator and power meter. The measurement technique consists simply of observing the power meter readings P_i, P_f without and with the two-port (or 'attenuator' i.e. Device Under Test) inserted and substituting these observations in eqn. 11.1. Provided that the generator and power meter are *matched*, the insertion loss will be equal to the attenuation. The failure to achieve this condition introduces an error which will be considered in greater

---

[17]See for example, F. L. WARNER, 'Microwave attenuation measurement,' (Peter Peregrinus Ltd, 1977)

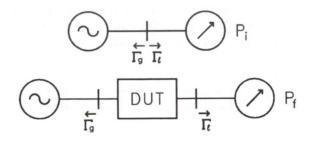

Fig. 11.1 'Power ratio' method for measuring attenuation

detail below. Despite its simplicity, this method, in conjunction with the *bolometric* power measurement technique, as described in Chapter 13, is among the most accurate of the existing methods for measuring small values of attenuation (10 dB or less). For larger values of attenuation, the dynamic range and operating power level considerations usually make alternative or *substitution* methods more attractive.

Substitution techniques

In the 'substitution' technique the attenuation of the unknown is *compared* with or measured against that of a *reference standard* whose attenuation characteristics are assumed to be known and whose operating frequency may equal that of the unknown, but alternatively may operate at some audio or 'intermediate' frequency or possibly even at DC. A simple example of the substitution technique is shown in Fig. 11.2 where the reference standard is connected in series with

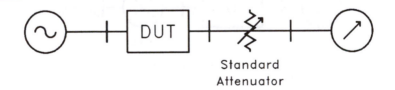

Standard
Attenuator

Fig. 11.2 Substitution technique for measurement of attenuation

the 'unknown' (or item to be calibrated) and operates at the same frequency. Here, the reference level indicated by the detector is first observed with the standard set to its minimum value. The unknown is then removed and the detector level returned to its prior value by an adjustment of the standard which then indicates the attenuation. (This assumes that the Γ_g and Γ_l, which obtain at the insertion point, vanish and combinations of isolators and tuning transformers are frequently required to meet this requirement.) Variants of this basic idea are shown in Figs. 11.3 and 11.4.

In Fig. 11.3, the amplitude of the microwave signal is modulated and the detector (ideally) provides an audio output which is proportional to the microwave signal level. As before, a reference

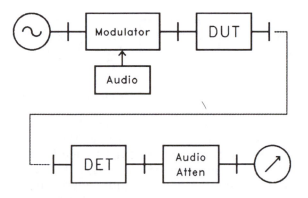

Fig. 11.3 Audio substitution technique for attenuation measurement

level is established with the unknown in place. Its attenuation may be determined from the change in the audio attenuator settings which are required to re-establish this reference value following the removal of the DUT.

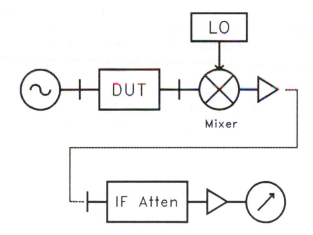

Fig. 11.4 Intermediate frequency technique for attenuation measurement

The operation of the circuit in Fig. 11.4 is similar, but the detector is now in the form of a mixer, which in turn requires a local oscillator. Ideally the mixer output, at the difference or intermediate frequency, will be proportional to the signal input. This permits the unknown to be measured against a reference attenuator which is operating at the difference or intermediate frequency, typically 30 MHz.

Parallel substitution

As an alternative to these examples of *series* substitution, one can also use *parallel* substitution, an example of which is shown in

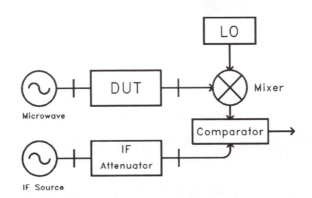

Fig. 11.5 'Parallel' substitution using IF techniques

Fig. 11.5. Here, an intermediate frequency, proportional to the signal level, is generated as in Fig. 11.4, but the detector is in the form of a comparator which alternately samples the mixer output and compares it with one derived from a separate source at the intermediate frequency and conveyed to the comparator via the intermediate frequency reference attenuator. This avoids certain potential problems in the stability of the detection circuit. Parallel substitution can also be used with microwave or audio frequency reference standards. The reference standards themselves will next be considered.

Microwave reference standards

Although many different forms of variable attenuators exist for the microwave frequencies, only one, or possibly two, are suitable as reference standards. The first of these is the rotary vane waveguide attenuator. As shown in Fig. 11.6, this device includes three basic components or sections in tandem. These include a rectangular to circular waveguide transition, a *rotating* centre section of circular waveguide (which includes a resistive vane in the plane defined by the longitudinal axis and a diameter) and a transition back to a

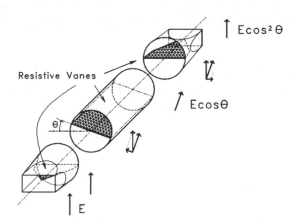

Fig. 11.6 Internal view of rotary vane attenuator

rectangular waveguide. A brief explanation of its operation will next be given, although it will be necessary to utilise the field description.

In the rectangular waveguide (TE_{10} mode) the (total) electric field **E** is transverse to the waveguide axis and normal to the broad or horizontal waveguide dimension.[18] In the centre section, the vane makes an (adjustable) angle θ with the horizontal. Upon entering this centre section, it is convenient to resolve **E** into two components which are, respectively, parallel and perpendicular to the vane. The former is absorbed by the vane, leaving a field of amplitude $E\cos\theta$ perpendicular to the vane and thus at the angle θ to the vertical. The rectangular to circular transitions also contain resistive vanes, which are parallel to the horizontal, thus the transition back to rectangular waveguide 'removes' the horizontal component of the electric field (of amplitude $E\sin\theta\cos\theta$) which emerges from the circular section, leaving a field amplitude of $E\cos^2\theta$ perpendicular to the horizontal. Since the power is proportional to E^2, the attenuation[19] is given by

$$Attenuation = -40\log_{10}(\cos\theta) \qquad (11.2)$$

---

[18]Its amplitude also varies sinusoidally across the broad dimension, although this detail is not required in what follows.

[19]More correctly, this is the *change* in attenuation as a function of vane angle. In addition there will be a residual component of attenuation even with $\theta = 0$. This, however, is of no concern in the present application.

Highly refined versions of this device have been developed at several of the national primary standards laboratories, including the National Bureau of Standards (now the National Institute of Standards and Technology) and the Royal Signals and Radar Establishment (RSRE) in England. Moreover, its performance compares favourably with the best of the alternative techniques.[20]   With the growing interest in broadband coaxial systems and the advent of the automated network analysers, however, the interest in this device has largely disappeared.

As an alternative to the rotary vane, the 'waveguide below cut-off' attenuator, which is described below as an 'intermediate frequency' reference standard, can also be used at microwave frequencies, although the practical problems are rather severe.

Intermediate and audio frequency reference standards

As noted in Chapter 2, the discussion to this point has been confined to that choice of dimensions and operating frequency such that the solution to the field equations is given by eqns. 2.1 and 2.2. In general, if the frequency of operation is increased, multimode propagation becomes possible and the solution becomes a superposition of terms, each of which satisfies eqns. 2.1 and 2.2. On the other hand, a single conductor waveguide (circular, rectangular etc.) is characterised by a 'cut-off' frequency, below which wave propagation no longer occurs and the fields are characterised by an exponential decay. The solution given by eqns. 2.1 and 2.2 still holds if $e^{\pm j\beta z}$ is replaced by $e^{\pm \alpha z}$, where α is a function of the cross-sectional dimensions. This result provides the basis for the 'piston attenuator' which, in a typical case, is comprised of a circular waveguide, an input network or 'launching coil' at one end and a moveable pick-up coil. As was the case with the rotary vane attenuator, this device has been the object of substantial refinements which include the use of a laser interferometer to measure the piston displacement etc.

Finally, the inductive voltage divider, or ratio transformer, has found extensive use as an audio frequency reference standard.

---

[20]The recent development of a second order theory and model which accounts for internal reflections, and other deviations from the ideal, now completes (in the author's judgment) the requirements for the projected use of the rotary vane attenuator as a *primary standard*. It is unfortunate that this theory was not available at an earlier date. See T. GULDBRANDSEN: 'Precision model for microwave rotary vane attenuator,' *IEEE Trans. Instru. & Meas.* IM-28, pp. 59-66, March 1979

Precision changes in power level

As an alternative to the microwave reference standard, it is possible to obtain precisely known changes in power level, (e.g. 6 dB) by multi-channel circuits. An example is shown in Fig. 11.7, which

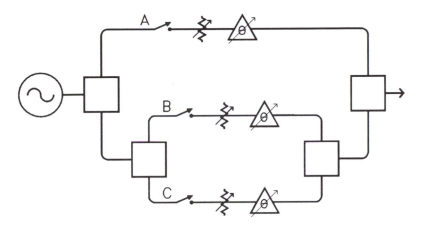

Fig. 11.7 Precision changes in power level (e.g. 6 dB) may be obtained by use of this circuit

provides three possible paths between the generator and detector, each of which contain a provision for adjusting the amplitude and phase of the transmitted signal, including a switch for no transmission. These paths are labelled *A*, *B* and *C* as shown. In operation the switches in paths *B* and *C* are first opened and the transmission via path *A* adjusted to some convenient value. The transmissions via paths *B* and *C* are then adjusted (in turn) for a detector null. At this point, if the switch in path *A* is opened, a 6 dB increase in the power transmitted from the source to detector will result if both of the switches *B* and *C* are closed as contrasted with either of them alone. A number of modifications of this basic concept are possible, which permit one to calibrate a variable attenuator at a number of discrete points. For proper operation, the dividing and combining networks must satisfy the 'multi-channel' isolation criteria as outlined in Chapter 6.

Accuracy considerations

Each of the techniques outlined above and the others which have not been included in this brief survey, have their own set of error sources, some of which are unique to the particular method[21], while others are shared by many or all of them. An example of the latter is the 'mismatch error' which is a result of the failure to satisfy the assumed $\Gamma_g = \Gamma_l = 0$ criteria at the insertion point.

Returning to Fig. 11.1, this error may be evaluated as follows: First, by use of eqn. 3.14, one has

$$P_i = \frac{|b_g|^2}{|1 - \Gamma_g\Gamma_l|^2}\left(1 - |\Gamma_l|^2\right) \tag{11.3}$$

Next, from eqn. 5.9, the parameters of the *equivalent* generator, b_{g2} and Γ_{g2}, which obtain at port 2 are given by

$$b_{g2} = \frac{b_g S_{21}}{1 - S_{11}\Gamma_g} \tag{11.4}$$

and

$$\Gamma_{g2} = \frac{(S_{12}S_{21} - S_{11}S_{22})\Gamma_g + S_{22}}{1 - S_{11}\Gamma_g} \tag{11.5}$$

These may now be substituted in eqn. 11.3, in place of b_g and Γ_g, to obtain P_f. Finally, one has

$$\frac{P_f}{P_i} = \frac{|S_{21}|^2\,|1 - \Gamma_g\Gamma_l|^2}{|(1 - S_{11}\Gamma_g)(1 - S_{22}\Gamma_l) - S_{12}S_{21}\Gamma_g\Gamma_l|^2} \tag{11.6}$$

Inspection of this result shows that if $\Gamma_g = \Gamma_l = 0$, then $P_f/P_i = |S_{21}|^2$, which confirms eqn. 5.10. When these criteria are not satisfied, it is a simple exercise to show that for small $|\Gamma_g|$ and small $|\Gamma_l|$, approximate limits for the error E are given by

$$E \approx \pm\Bigl(2\,|S_{11}\Gamma_g| + 2\,|S_{22}\Gamma_l| + 3\,|S_{11}\Gamma_g|^2 + 3\,|S_{22}\Gamma_l|^2$$
$$+ 2\bigl[1 + |S_{12}S_{21}| + 2\,|S_{11}S_{22}|\bigr]\,|\Gamma_g\Gamma_l|\Bigr) \tag{11.7}$$

or in decibels

$$E_{dB} \approx \pm 10\log_{10}\bigl(1 + |E|\bigr) \tag{11.8}$$

---

[21]For a more complete description, see the reference cited in Footnote 17.

Chapter 12
The microwave reflectometer

In one form or another, the field of microwave impedance measurement is dominated today by the *reflectometer*. This includes, as special cases, the 'scalar' and 'vector' network analysers, the 'six-port' etc. By analogy to low-frequency methods, the reflectometer is closely related to the impedance 'bridge' but with one important difference. At low-frequencies, one or more of the bridge parameters (or arms) are usually adjustable and the unknown is measured in terms of these and at the 'null' condition. Although the counterpart can be constructed at microwave frequencies as well[22], the associated problem of implementing the equivalent of the adjustable arms, for which *known* values of impedance are required, is much more difficult. Instead, one uses a 'bridge' of *fixed* parameters and the result of interest is obtained in terms of these parameters *and* the 'off-null' signal response.

Elementary considerations

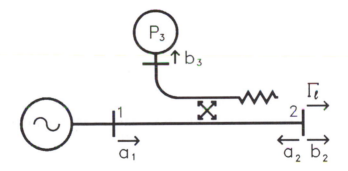

Fig. 12.1 Elementary reflectometer

In Fig. 12.1 one has a reflectometer in its most elementary form. The sidearm response is given by

$$b_3 = S_{31}a_1 + S_{32}a_2 + S_{33}a_3 \tag{12.1}$$

[22]See, for example: M. CHODROW, E. L. GINZTON, and F. KANE: 'A microwave impedance bridge,' *Proc. IRE*, 37, No. 9, pp. 634-639, June, 1949

For convenience an ideal coupler, for which $S_{31} = 0$, and a matched detector, for which $a_3 = 0$, is assumed. Under these conditions, and if $b_3 = 0$, one concludes that $a_2 = 0$, which will be true for a matched load. The technique thus permits recognition or adjustment of a termination for an impedance match. To continue, assuming the detector responds only to power one has

$$P_3 = |S_{32}|^2 |a_2|^2 \qquad (12.2)$$

while

$$|a_2|^2 = |b_2|^2 |\Gamma_l|^2 \qquad (12.3)$$

thus, adding the subscript '*l*' to P_3 for 'load'

$$P_{3l} = |b_2|^2 |S_{32}|^2 |\Gamma_l|^2 \qquad (12.4)$$

and P_{3l} is proportional to $|\Gamma_l|^2$ *provided* that $|b_2|^2$ is constant. Assuming this to be the case, the proportionality factor can be evaluated by observing the response to a termination of known reflection, a convenient example being a short for which $|\Gamma_l| = 1$. If the reflectometer response to the short is denoted by P_{3s}, one has

$$P_{3s} = |b_2|^2 |S_{32}|^2 \qquad (12.5)$$

and finally

$$|\Gamma_l|^2 = \frac{P_{3l}}{P_{3s}} \qquad (12.6)$$

In achieving this result, however, three major assumptions have been made. These include

(1) Ideal coupler

(2) 'Constant' value of $|b_2|$

(3) Matched detector on arm 3

In practise, none of these criteria are satisfied. The remainder of the chapter will (1) describe the operation from a more general viewpoint, (2) develop certain 'tuning' techniques for more nearly realising the 'ideal' response, (3) outline and discuss the relative merit of several alternative reflectometer configurations and (4) make an assessment of the measurement errors which result from a failure to completely realise this objective. Although the flexibility of the method may be greatly enhanced by the addition of phase response in the detectors, such as is found in the 'vector' network analysers, this also makes a substantial addition to the complexity of the detection system. This chapter will explore the question of what can be achieved by the use of detectors whose response is limited to that of providing amplitude (or power) response only. (The use of phase response will be considered in the chapters which follow.)

General theory

In keeping with the arguments given in the latter part of Chapter 8, the detector reflection can be 'absorbed' by the coupler parameters. The detector response is then given by eqn. 8.13 which may also be written

$$b_3 = b_2 \left\{ \left[S_{32} - \frac{S_{31}S_{22}}{S_{21}} \right] \Gamma_l + \frac{S_{31}}{S_{21}} \right\} \tag{12.7}$$

and where the nonideal coupler and detector characteristics are reflected by the terms S_{31} and S_{22}. It remains to consider the effect of a nonconstant value of $|b_2|$.

Returning to eqn. 5.9, one has an explicit expression for b_2 as a function of the source parameters b_g, Γ_g and the coupler scattering coefficients. (Note that although this result was originally derived for a two-port, the same arguments hold for a three-port if the third port is terminated by a match. The specific details, however, are unnecessary in the present context.) As an alternative to eqn. 5.9, it is more convenient to simply write

$$b_2 = b_{g2} + a_2 \Gamma_{g2} \tag{12.8}$$

where the added subscript '2' indicates that the parameters are those which obtain at the coupler output port. Thus from eqn. 3.11 one has

$$b_2 = \frac{b_{g2}}{1 - \Gamma_{g2}\Gamma_l} \tag{12.9}$$

and the substitution of this in eqn. 12.7 gives

$$b_3 = b_{g2} \frac{\left[S_{32} - \dfrac{S_{31}S_{22}}{S_{21}} \right] \Gamma_l + \dfrac{S_{31}}{S_{21}}}{-\Gamma_{g2}\Gamma_l + 1} \qquad (12.10)$$

which is the desired result.

Although the actual generator parameters have been absorbed by b_{g2} and Γ_{g2}, there is still a strong implicit functional dependence. In particular, b_{g2} is proportional to b_g. Thus, one is led to the (probably already obvious) conclusion that the response of b_3 is directly proportional to b_g and, to the extent that this is unstable, this is the equivalent of a measurement error in b_3 or P_3. Although there may be applications where this can be tolerated, the detector performance is usually substantially better than that of an unstabilised source and it is generally desirable to eliminate the potential error due to amplitude instability in the generator.

This may be conveniently done by the addition of the amplitude stabilisation circuit as described in Chapter 6. At this time, either of two points of view may be adopted: (1) the amplitude stabiliser and reflectometer circuits could be maintained as separate entities or (2) the two may be merged such that the 'four-port' reflectometer results,

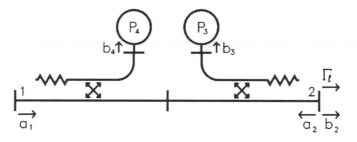

Fig. 12.2 Four-port reflectometer

as shown in Fig. 12.2. Although for some problems either point of view is perhaps equally satisfactory, there are others where the second provides more flexibility.

The dependence on the generator may now be eliminated by taking the ratio of the responses of the two detectors. The question of whether to actually add the feedback, and thus stabilise the output, becomes a separate issue. Moreover, the anticipated addition of phase response in the detectors is easier in this second context. Although the four-port reflectometer will be the primary subject in the treatment which follows, it should be recognised that the signal sources which are available today frequently include internal stabilisation and which is often beyond the convenient reach of the metrologist. In some cases this may eliminate the need for the second coupler and the first point of view may prove of substantial value.

Returning to eqn. 8.13, if the subscript 4 is substituted for 3, the resulting equation may be written

$$b_2 = \frac{\dfrac{S_{21}}{S_{41}} b_4}{-\left[S_{22} - \dfrac{S_{42}S_{21}}{S_{41}}\right]\Gamma_l + 1} \tag{12.11}$$

and by comparison with eqn. 12.9, the parameters of the resulting equivalent generator now become

$$b_{g2} = b_4 S_{21}/S_{41} \tag{12.12}$$

$$\Gamma_{g2} = (S_{22} - S_{42}S_{21}/S_{41}) \tag{12.13}$$

which (after substituting '3' for '4') is also in agreement with eqn. 6.5 if $\Gamma_d = 0$. After making this substitution, eqn. 12.10 becomes

$$\frac{b_3}{b_4} = \frac{\left[\dfrac{S_{32}S_{21}}{S_{41}} - \dfrac{S_{31}S_{22}}{S_{41}}\right]\Gamma_l + \dfrac{S_{31}}{S_{41}}}{-\left[S_{22} - \dfrac{S_{42}S_{21}}{S_{41}}\right]\Gamma_l + 1} \tag{12.14}$$

and finally

$$\frac{b_3}{b_4} = w = \frac{A\Gamma_l + B}{C\Gamma_l + 1} \tag{12.15}$$

where

$$A = (S_{32}S_{21} - S_{31}S_{22})/S_{41}$$

$$B = S_{31}/S_{41} \tag{12.16}$$

$$C = -(S_{22} - S_{42}S_{21}/S_{41})$$

Although most of the foregoing could have been written as an immediate consequence of eqn. 8.1, the objective has been that of providing additional insight into the operation.

'Ideal' reflectometer

Returning to eqn. 12.15, an 'ideal' reflectometer results if $B = C = 0$, such that

$$P_3/P_4 = |A|^2|\Gamma_l|^2 \tag{12.17}$$

(One might also wish to add the further qualification that $|A| = 1$. This calls for $|S_{32}S_{21}| = |S_{41}|$. Although this condition may be approximated by certain configurations, the ease with which departures from it may be corrected leaves little incentive for trying to achieve this result.)   For an ideal pair of couplers, $S_{31} = S_{22} = S_{42} = 0$ which, from inspection of eqn. 12.16, provides the desired result. More generally, B or S_{31}/S_{41} is the ratio of the signal amplitudes which are transmitted from port 1 to ports 3 and 4, respectively, given that port 2 is terminated by a match. This term is closely related to the coupler directivity. The expression for B may now be written

$$B = \frac{S_{31}S_{32}}{S_{32}S_{41}} \tag{12.18}$$

and where the directivity is given by

$$\text{Directivity} = 20\log_{10}(S_{31}/S_{32})^{-1} \tag{12.19}$$

and $|S_{32}/S_{41}|$ is of the order of unity if the parameters for the two couplers are similar or equal.

With regard to the factor C, whose presence in eqn. 12.15 also represents a deviation from the 'ideal', this is just the negative of the reflection coefficient for the equivalent generator which now obtains at port 2. Although primarily a function of the coupler which samples the incident wave, it is important to note that the equivalent generator, which would otherwise exist, has here been modified by the addition of a second coupler which samples the reverse wave. In other words, and from the viewpoint of the stabilisation circuit in Chapter 6, a 'two-port' has been added at its output and the equivalent generator parameters are modified as described in Chapter 5. Ideally, the coupler or 'two-port' is matched or free of reflection, while its 'attenuation' is a function of the coupling ratio. In a typical setup, this attenuation is small, but in some applications may be substantial. In any case, the resultant modification of the stabilised source may be described by eqn. 5.9, although the added detail is not necessary in the present context. As a counterpart to eqn. 12.6, one has

$$|\Gamma_l|^2 = \frac{P_{3l}}{P_{4l}} \cdot \frac{P_{4s}}{P_{3s}} \qquad (12.20)$$

The 'tuned' reflectometer

In general terms, the magnitudes of B and C and thus the errors in using eqn. 12.20 to obtain $|\Gamma_l|$ will be dependent upon how closely the design objectives are realised by the hardware. In contrast with the slotted line, a substantial improvement in the performance of this technique may be realised by methods which are easily within the scope of the metrologist. Indeed, for more than a decade, but prior to the advent of digital and automated methods, these techniques provided much of the basis for precision microwave metrology. Although these methods have now been largely replaced by software corrections, which are implicitly or explicitly based on phase information, a brief summary of the tuned reflectometer will be included in the interest of completeness.

In general, the technique calls for the addition of adjustable tuning transformers, T_X and T_Y, as shown in Fig. 12.3 and the use of

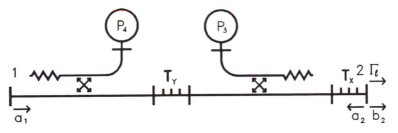

Fig. 12.3 Addition of tuners to create the 'tuned' reflectometer

strongly and weakly reflecting sliding terminations, whose role will be explained below.

First, with respect to the parameter B, and referring to Fig. 12.4, this can be made to vanish by the simple expedient of adjusting T_X such that b_3 vanishes with a matched termination on

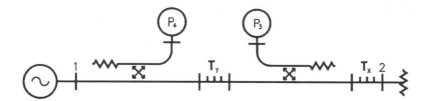

Fig. 12.4 Adjust T_X for a null, $(P_3=0)$ with a <u>matched</u> termination $(\Gamma_I=0)$ on arm 2

arm 2. (Note that, by definition, this requires that $S_{31} = 0$). Next the parameter C (as noted above) is the negative of the source impedance for the equivalent generator which is obtained at port 2 under the condition that P_4 is constant. Moreover, as explained in Chapter 6, this is also the impedance which is observed at port 2 after port 1 has been terminated in such a way that $P_4 = 0$. Thus referring next to

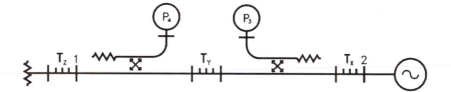

Fig. 12.5 Adjust T_Z for a null $(P_4=0)$ then adjust T_Y such that an impedance match is obtained at port 2

Fig. 12.5, the signal source is connected to port 2, a termination and tuner T_Z are connected to port 1, and the latter adjusted such that $P_4 = 0$. Tuner T_Y is then adjusted such that the reflection coefficient *looking into* port 2 now vanishes.[23] From the description of the directional coupler operation given in Chapter 6, it may be recognised that the condition for a null in P_4 is independent of the adjustment

---

[23]This may be realised, for example, by the previously described method of recognising an impedance match

of T_Y while the impedance presented at port 2 will certainly depend on its value. As further explained in Chapter 6, T_X could alternatively be adjusted for an impedance match at port 2 which would also provide $C = 0$. This, however, would destroy the earlier condition $B = 0$. What is needed is an alternative tuner and location which will not interfere with, but instead preserve the $B = 0$ adjustment. This is provided by the tuner T_Y. The insensitivity of the $B = 0$ (or $P_3 = 0$) condition to its operation may be recognised by interpreting T_Y as part of the 'signal source characteristics' in Fig. 12.4. As an alternative to the given location, one could also incorporate the tuner as part of the detector on arm 3. Depending on the coupling ratios and configuration, there are times when this is may be preferred.

In principle the desired condition, $B = C = 0$, may be realised by the given procedure. As a practical matter, the accuracy of the T_X adjustment will be limited by deviations of the termination on port 2 from the assumed impedance match and, in the case of T_Y, by one's inability to recognise when the desired impedance match has been achieved. Moreover, the described procedure calls for removal of the generator from port 1, reconnecting it to port 2 etc. These problems are avoided by the sliding termination techniques which will be next described.

Consider first the reflectometer response to the weakly reflecting sliding termination. If it were totally nonreflecting, the response would be $P_3/P_4 = |B|^2$. In general, the reflection coefficient of a sliding termination is described by

$$\Gamma_l = re^{j\theta} \tag{12.21}$$

where θ is a function of the longitudinal position. For a 'weakly' reflecting termination r may be 0.01 or smaller, while for a sliding 'short' r should be close to unity. As described in Chapter 9 and referring to eqns. 9.4 and 9.5, the locus of values which are generated in the w-plane, in response to the motion, will be a circle of radius R and centre R_c which are given by

$$R = \frac{|A - BC| \, |\Gamma_l|}{1 - |C|^2 |\Gamma_l|^2} \tag{12.22}$$

and

$$R_c = \frac{B - AC^* |\Gamma_l|^2}{1 - |C|^2 |\Gamma_l|^2} \tag{12.23}$$

To be more specific, and referring to eqn. 9.3, w is given by the sum of two terms, one is a constant (R_c), the other is of constant amplitude but of variable phase (R). If one now seeks the condition that the response is constant, with respect to the motion of the sliding load, one concludes that either $R = 0$ or $R_c = 0$. If the former is the case, the sliding termination is perfectly matched and it is only necessary to adjust for a null output which leads to $R_c = 0$ and thus $B = 0$. In the general case, $R \neq 0$ and the adjustment for a constant output requires $R_c = 0$ which implies

$$B = AC^* |\Gamma_l|^2 \qquad (12.24)$$

By hypothesis $|\Gamma_l|$ is small. Moreover, eqn. 12.24 requires that $|B|$ is smaller than $|AC|$ by the factor $|\Gamma_l|^2$. With the sliding termination replaced by the sliding short, the same arguments may be repeated, except that the objective of the adjustment is now that $C \to 0$. Thus eqn. 12.24 becomes

$$AC^* = B \qquad (12.25)$$

and $|AC|$ has been reduced to the size of $|B|$.

Although in principle this calls for an iterative procedure, in practise values of $|\Gamma_l|$ as small as 0.01 or less are not uncommon. Thus the desired result can be usually achieved by a single adjustment each of T_X and T_Y. Once this has been done, the operation calls for observing the response with a short, $|\Gamma_l| = 1$, which then determines $|A|$ and the response of interest is again given by eqn. 12.20.

Alternative reflectometer configurations

Although the coupler arrangement shown in Fig. 12.2 is the most common, a variety of other configurations are not only possible but in some cases preferred. In Fig. 12.6, for example, the order of the couplers has been reversed. Here, the same basic tuning methods may be employed but with due consideration for their potential interaction as described below. In order to better appreciate the constraints involved, it may prove useful to derive a slightly more general form of eqn. 12.15. Starting with eqns. 7.5 and 7.6 or indeed from eqn. 8.1 where $i = 3, 4$ and after a change in notation[24] ($A = A_3$, $B = B_3$, $C = A_4$, $D = B_4$) one has

---

[24]The definitions given by Equations 12.16 are no longer applicable.

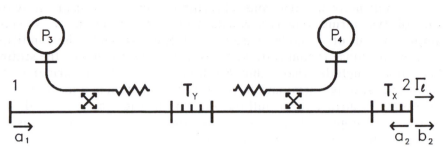

Fig. 12.6 Alternative reflectometer configuration

$$P_3 = |Aa_2 + Bb_2|^2 \tag{12.26}$$

and

$$P_4 = |Ca_2 + Db_2|^2 \tag{12.27}$$

The first point to be made is that, for the configuration of Fig. 12.3, A and B are functions only of the coupler on the right. Although this follows from the arguments given in Chapter 8, for example and where the coupler on the left may be considered to be part of the 'source', an alternative proof will be given.

From eqn. 8.1, and the change in variables introduced above, the solution to the scattering equations of the complete circuit (including *both* couplers in Fig. 12.3) may be written in the general form

$$b_3 = Aa_2 + Bb_2 \tag{12.28}$$

Next, let an auxiliary terminal plane be inserted between these couplers, possibly, but not necessarily, in conjunction with the tuner T_Y. In terms of the subassembly thus defined, one can also write

$$b_3 = A'a_2 + B'b_2 \tag{12.29}$$

where A' and B' are functions *only* of the remaining coupler or subassembly. After subtracting eqn. 12.29 from eqn. 12.28 one has

$$(A-A')a_2 + (B-B')b_2 = 0 \tag{12.30}$$

thus a nontrivial relationship exists between b_3, a_2 and b_2 if, and only if, $A = A'$ and $B = B'$. This is the result of interest.

To continue with the discussion of alternative coupler configurations and of Fig. 12.3, the parameters C and D will be functions of *both* couplers, since the coupler on the right may be regarded as an extension of the output arm or port 2 of the one on the left. If, however, as shown in Fig. 12.6, the order of the couplers has been reversed, the converse is true. In particular, C and D now depend only on the coupler on the right, whereas A and B are functions of both.

The ratio of eqn. 12.26 to eqn. 12.27 may be written

$$\frac{P_3}{P_4} = \frac{|A|^2 |\Gamma_l + (B/A)|^2}{|D|^2 |(C/D)\Gamma_l + 1|^2} \tag{12.31}$$

which may be regarded as an alternative, albeit slightly more general form of eqn. 12.15. Here the ratios B/A and C/D are, in turn, independent of the coupler on the left in Figs. 12.3 and 12.6 (including T_Y and its adjustment).

With the foregoing as background, it is now possible to make some observations as to the relative merits of the two configurations. With respect to impedance measurement, it will be shown (in the following section) that the error for small values of reflection will be dominated by the nonzero value of B or more specifically B/A. Thus the insensitivity of this parameter to the second coupler (and the potential instability associated therewith) is an argument for the configuration in Fig. 12.3. On the other hand, in certain power measurement applications, yet to be described, the coupler will emerge as a calibrated signal source. Here, the insensitivity of the equivalent Γ_g and b_g to the second coupler, which is provided by the configuration of Fig. 12.6 is an argument in favour of its use.

In either case, and as already noted, the same sliding load and sliding short techniques may be used to effect the adjustment of T_X and T_Y, although some differences may be noted. For this alternative configuration, it is probably desirable to begin with the sliding short and adjust T_Y such that eqn. 12.25 is satisfied. Following this, the adjustment of T_X is effected with the help of the sliding load as before. (Note that, given the proper adjustment of T_X, this now simultaneously drives *both* B and C to zero.) In principle, if the sliding short is truly of unit reflection *and* if the tuner T_X is free of dissipation, the two tuning adjustments will be independent of each other. This may be recognised by noting that, under the assumed conditions, and in which energy dissipation has been excluded, the magnitude of the reflection coefficient presented by the tuner and sliding short combination must be unity. *Only the phase of the reflection is affected by the tuner adjustment.* Thus, if the ratio P_3/P_4 is constant with respect to the short motion, for one adjustment of T_X, the same must be true for all adjustments. In practise, where these idealisations are never completely satisfied, there will be some, and probably more, interaction between the adjustments than for the configuration of Fig. 12.3. This may constitute an argument for the use of the latter.

Although it is possible to identify a number of other configurations, only one more will be noted here. In some applications it is desired to achieve the maximum possible signal, P_3,

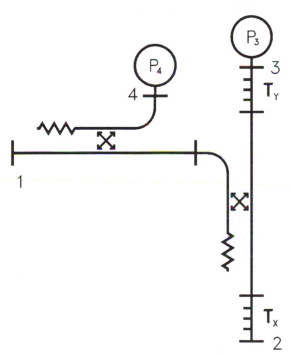

Fig. 12.7 Another reflectometer configuration

for a given magnitude $|b_2|$ of the signal impressed on Γ_l. In the configuration of Fig. 12.3, the reflected signal suffers an attenuation due to the coupling ratio. This is usually a minimum of 3 dB, and is often substantially larger. In this context the arrangement of Fig. 12.7 is useful. Here the output of the first coupler feeds the sidearm of the second. In this case, if the coupling ratio is 10 dB, for example, 90% of the signal power reflected from the termination on port 2 will be available to the detector on port 3. This also represents a case where the choice of tuner location for T_Y is probably ahead of the detector on arm 3 as shown in Fig. 12.7. The basis for this may be recognised by recalling the requirement for an impedance match at port 2 after port 1 has been terminated such that $P_4 = 0$. If, instead of the position shown, the tuner is inserted between the couplers, it is also 'decoupled' from the impedance which obtains at port 2 by the coupling ratio, thus a much larger adjustment would be required to yield the desired effect. This, at best, would probably prove more frequency sensitive and might not even be realisable within the range of available adjustment.

Assessment of error due to nonideal network

Although one of the major uses of the sliding termination has been in the context of tuning adjustments, it is also possible to use the observed response to these devices to evaluate certain sources of error. In a scalar network analyser, for example, tuning techniques are usually incompatible with the desired broadband operation and the failure of the measurement network, or test set, to satisfy the $B = C = 0$ conditions, as identified above, represents a source of error. Since the failure to satisfy these criteria is reflected in the (nonconstant) system response to the sliding terminations, the objective will be to express the error in an measurement of an unknown reflection Γ_u in terms of these observations. This exercise will provide a further demonstration of the utility of the sliding termination techniques.

Starting with eqn. 12.31, a change of variables will prove useful. Let $b = B/A$, $c = C/D$ and $a = A/D$ so that

$$\frac{P_3}{P_4} = |w|^2 = |a|^2 \frac{|\Gamma_u + b|^2}{|c\Gamma_u + 1|^2} \qquad (12.32)$$

By hypothesis, $|b|$ and $|c|$ are small and $|\Gamma_l| \le 1$ so that approximately

$$\frac{P_3}{P_4} = |a|^2 \left| (\Gamma_u + b)(1 - c\Gamma_u + c^2\Gamma_u^2 \cdots) \right|^2 \qquad (12.33)$$

which may be further expanded to yield

$$\frac{P_3}{P_4} = |a|^2 \left(|\Gamma_u|^2 + b^*\Gamma_u + b\Gamma_u^* + |b|^2 \right) \times$$
$$\left(1 - c\Gamma_u - c^*\Gamma_u^* + |c\Gamma_u|^2 + \left[c\Gamma_u \right]^2 + \left[c^*\Gamma_u^* \right]^2 + \cdots \right) \qquad (12.34)$$

In terms of the parameter set introduced above, eqns. 12.22 and 12.23 become

$$R = \frac{|a|\,|1 - bc|\,|\Gamma_l|}{1 - |c|^2\,|\Gamma_l|^2} \qquad (12.35)$$

and

$$R_c = \frac{a\left[b - c^*\,|\Gamma_l|^2 \right]}{1 - |c|^2\,|\Gamma_l|^2} \qquad (12.36)$$

For a weakly reflecting sliding load, $|\Gamma_l| \ll 1$. Adding the subscript 'l', eqns. 12.35 and 12.36 become

$$R_l \approx |a|\,|\Gamma_l| \qquad (12.37)$$

$$R_{cl} \approx ab \qquad (12.38)$$

since by hypothesis $|\Gamma_l|$ and $|bc|$ are small. Using the subscript 's' to represent the sliding short, and assuming $|\Gamma_s| \approx 1$, one has

$$R_s \approx |a| \qquad (12.39)$$

and

$$R_{cs} \approx a(b - c^*) \qquad (12.40)$$

since $|c|^2$ and $|bc|$ are small with respect to unity.

Let

$$d = b - c^* \qquad (12.41)$$

which is thus proportional to R_{cs}. After solving eqn. 12.41 for c, substituting in eqn. 12.34, and collecting terms, one has to the first order in b and d,

$$\frac{P_3}{P_4} \approx |a|^2 \left[|\Gamma_u|^2 + \left[b\Gamma_u^* + b^*\Gamma_u \right] \left(1 - |\Gamma_u|^2 \right) \right.$$
$$\left. + \left[d\Gamma_u^* + d^*\Gamma_u \right] |\Gamma_u|^2 + |b|^2 \right] \qquad (12.42)$$

where the second order term, $|b|^2$, has also been retained since this characterises the error as $\Gamma_u \to 0$.

To continue, eqns. 12.38 $\cdots$ 12.41 may be combined to yield

$$|b| \approx \frac{|R_{cl}|}{R_s} \qquad (12.43)$$

and

$$|d| \approx \frac{|R_{cs}|}{R_s} \qquad (12.44)$$

which provide (approximate) values for $|b|$ and $|d|$ in terms of the observable parameters $|R_{cl}|$, $|R_{cs}|$ and R_s. In general, the accompanying arguments are unknown thus, assuming a worst case phase relationship in eqn. 12.42, one has

$$\frac{P_3}{P_4} \approx |a\Gamma_u|^2 \left(1 \pm 2 \left| \frac{R_{cl}}{R_s} \right| \frac{1 - |\Gamma_u|^2}{|\Gamma_u|} \right.$$
$$\left. \pm 2 \left| \frac{R_{cs}}{R_s} \right| |\Gamma_u| + \left| \frac{R_{cl}}{R_s} \right|^2 \frac{1}{|\Gamma_u|^2} \right) \qquad (12.45)$$

which is the result of interest.

If $|a|$ is known, $|\Gamma_u| \gg |R_{cl}/R_s|$ and $|\Gamma_u| \gg |R_{cs}/R_s|$, then eqn. 12.45 implicitly describes the error in a measurement of $|\Gamma_u|$ due to nonzero values of R_{cl} and R_{cs}. In particular one may obtain limits for $|\Gamma_u|$ by use of the $\pm$ signs. On the other hand, as may be confirmed by eqn. 12.32, if $|\Gamma_u| \to 0$, $(P_3/P_4) \to |a|^2$, and the response no longer depends on Γ_u. In the interval where $|\Gamma_u| \approx |R_{cl}/R_s|$, both the first and second order terms in $|R_{cl}/R_s|$ are important and an aid in interpreting eqn. 12.45 is obtained as follows. By use of eqn. 9.3, and referring to Fig. 12.8, for small values of $|\Gamma_u|$ the response w will be given approximately by the *vector* sum of R_c (or R_{cl}) and $a\Gamma_u$, but where the angle between them is unknown. By inspection

$$| \, |R_c| - |w| \, | \; \leq \; |a\Gamma_u| \; \leq \; |R_c| + |w| \qquad (12.46)$$

In general there are two cases where $|\Gamma_u|$ is 'well defined': $|w| \ll |R_c|$ and $|w| \gg |R_c|$. The first of these will only obtain, however, if Γ_u happens to very nearly equal the negative of R_{cl}/a and

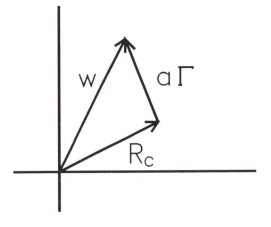

Fig. 12.8 *Relationship between w, R_c, and aΓ*

is therefore of only limited interest. Thus, one is lead to consider the alternative where $|w/a|$ and thus $|\Gamma_u|$ is substantially (perhaps a factor of 10 or more) larger than $|R_{cl}/a|$ as has been considered above. In this case the term which contains the factor $|R_{cl}/R_s|^2$ can be neglected since the error is now dominated by the two terms in $|R_{cl}/R_s|$ and $|R_{cs}/R_s|$ in eqn. 12.45. If $|\Gamma_u|$ is still 'small', say of the order of 0.1, the major source of error will now be due primarily to the first if $|R_{cl}|$ and $|R_{cs}|$ are nominally equal. For example, if $|R_{cl}/R_s| = 0.01$ and $|\Gamma_u| = 0.1$, one has a nominal 20% error due to the first and 0.2% from the second. As $|\Gamma_u|$ increases, the error from the first term diminishes and vanishes in the limit as $|\Gamma_u| \rightarrow 1$. Thus for large $|\Gamma_u|$, the error is due primarily to the second term.

For completeness, it is desirable to take a more careful look at the approximate relationship, eqn. 12.40. By use of techniques similar to those employed with eqn. 12.32, one can show

$$R_s \approx |a\Gamma_s| \left(1 + \left| \frac{R_{cs}}{R_s} \right|^2 - \frac{R_{cl}R_{cs}^* + R_{cl}^* R_{cs}}{2R_s^2} + \cdots \right)$$

$$(12.47)$$

thus the error in eqn. 12.40 is of second order provided that $|\Gamma_s| \approx 1$. In general, one has two choices for assigning a value to $|a|$. First, as suggested above, one can use R_s in which case $|a|$ will be in error by the factor $(1 - |\Gamma_s|^2)$. The other alternative is to use eqn. 12.45 in conjunction with a fixed short, for which the deviation from unity should be negligible. In this case, $|a|$ may be in error by as much as $\pm |a| |R_{cs}/R_s|$. Ordinarily, the choice should be made in favour of the smaller error.

Summary

As noted in an earlier paragraph, the technology described above at one time represented a highly developed art and provided much of the basis for precision microwave metrology. Although failing to yield phase information, the need for this was in some cases circumvented by 'generalised reflectometer techniques' which specifically addressed certain problems of major interest, particularly in the area of power measurement.[25] At best, however, these methods were both frequency sensitive and time consuming. With the increasing interest in broadband systems, and the advent of digital technology, these methods have to a large degree been replaced by others which are more amenable to automation. In the process, a major shift in measurement strategy has also occurred. In particular, the foregoing tuning methods were largely developed in an era where the key to better accuracy was an improved item of hardware. As such, they represent a set of highly developed techniques for making in-situ adjustments of the hardware parameters such that the 'ideal' response is more nearly realised. By contrast, the more recent strategy, as reflected by the 'vector' network analyser, for example, is that of specifically identifying and characterising the hardware imperfections such that they may be eliminated by software corrections. Although much of this earlier technology is still embodied in the 'scalar' network analyser, in other cases the shift in emphasis has been from trying to build a better piece of hardware, to 'smarter' use of that already in existence.

---

[25]Examples of these will be given in Chapter 15

Chapter 13

Fundamentals of
Precision power measurement

At microwave frequencies the lower frequency concepts of voltage and current loose most of their practical significance and, to a substantial degree, their role is replaced by that of *power*. Loosely speaking, the question of 'how much' signal is present is now answered by a power rather than a voltage or current measurement. Given this analogy (albeit a rather crude one) it is important to note certain distinctions in order to avoid the potential trap of carrying the analogy too far. An 'ideal' meter would permit one to measure power flow in a given waveguide, possibly in both directions, without extracting any of the attendant energy. Although it is possible to approach this ideal in high power systems, at lower power levels and in contrast to a voltage or current meter, the measuring instrument typically absorbs the power being measured.

The importance of power measurements in assessing the performance of a microwave system may be easily recognised since it is power which 'does the work' or in the communications environment, 'carries the information'. The amount of power needed for a communications application is dictated, in turn, by the required rate of information transfer, carrier frequency, noise thresholds etc. By tradition the field of microwave metrology is divided into the areas of power, attenuation and impedance measurement. The primary role of attenuation measurement, however, and to a substantial degree that of impedance as well, is ultimately in support of power measurement.

The phenomena which have been investigated and used for the determination of power at microwave frequencies include mechanical forces or torque, field interaction with an electron beam, the Hall effect and a number of others.[26] Although differing degrees of

---

[26]Where the ultimate in accuracy is not required, extensive use is made of diodes which at low power levels provide a current which is *approximately* proportional to the power level. They have found only a limited application to *precision* power measurements, however.

success have been reported in these efforts, the primary interest has largely reverted to the simple expedient of converting the energy to heat which then becomes the measurement objective. As noted above, this frequently means that, at low power levels at least, the power meter also absorbs the energy in the process of its measurement. At higher power levels, and to some extent at lower ones, the available techniques also include some for which only a fraction of the total energy is thus disposed of. It is not the intent, however, to provide a survey of the available methods or associated instrumentation, but rather to describe their application to power measurement problems and identify some common sources of error. Because of the major role which it has played in precision metrology, it will be convenient to focus on the bolometric technique, identify its major sources of error (most of which have their counterpart in other methods) and outline the methods which have been devised to evaluate them.

As used in the context of microwave metrology, the 'bolometer' is a temperature sensitive resistor. The associated 'bolometric' technique is one in which a determination of microwave power is achieved by comparing its heating effect on the bolometer with that of a measured amount of low-frequency power.

The bolometer may consist of a fine platinum or 'Wollaston' wire or a restive film which has been deposited on a suitable substrate, such as mica. These are characterised by a positive temperature coefficient and carry the name, 'barretter'. Alternatively, the bolometer may be formed from a small 'bead' of

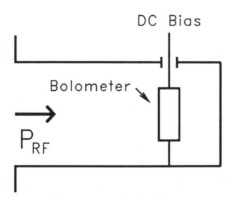

Fig. 13.1 Bolometer mount

semi-conducting material and exhibit a negative temperature coefficient (thermistor). As a result of their greater immunity to burnout from overload, larger temperature coefficient of resistivity and better mechanical stability, the existing art is largely dominated by the thermistor types, although the barretter is still of interest in certain metrology applications. As shown in Fig. 13.1, a companion structure or 'bolometer mount' is essential to the application. This includes provision for coupling the bolometer to both the microwave (or 'RF') energy and to a low-frequency (preferably DC) source of bias power. Ideally, the bolometer-bolometer-mount combination terminates the transmission line in its characteristic impedance.

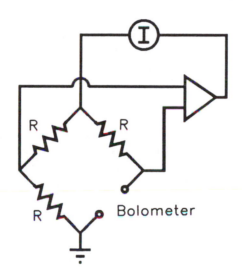

Fig. 13.2 Bolometer bridge

Historically, as shown in Fig. 13.2, the bolometer forms one arm of a Wheatstone bridge. This provides both the source of bias power and a method (via the bridge null) for recognising when the bolometer is at some predetermined value of operating resistance, R, (typically 200Ω in waveguide systems). More recently, improved performance has been achieved by elimination of the bridge which has been replaced by a single resistor and an additional operational amplifier.[27]   Returning to Fig. 13.2, a common arrangement includes a DC bias source and equal bridge arms of resistance R. For convenience, it is further assumed that feedback is provided which adjusts the total bridge current I, as required, to establish and maintain the bridge balance. Since the total bridge resistance (as 'seen' by the source) is also R, the *total* power dissipated therein is just I^2R and, by symmetry, 1/4 of this occurs in the bolometer. In the absence of microwave energy, and at bridge balance, the power dissipation P_{dc} in the bolometer is thus given by

$$P_{dc} = I^2R/4 \qquad (13.1)$$

If microwave power is now applied, this will also contribute to the bolometer heating and, if bridge balance is to be maintained, it will be necessary to reduce the DC bias power by a (nominally!) equal amount.

The *change* in DC bias power, P_{bolo}, which is required to achieve this result is given by

$$P_{bolo} = \frac{R}{4}\left[I_1{}^2 - I_2{}^2\right]$$

The use of this result as a measure or indication of microwave power is accompanied by several sources of error. These include:

(1) Instrumentation error

(2) DC–RF substitution error
(or just substitution error)

(3) Bolometer mount efficiency

---

[27]N. T. LARSEN, 'A new self-balancing DC-substitution RF power meter', *IEEE Trans. Instrum. Meas.*, IM-25, pp. 343-347, Dec. 1976

To these is sometimes added a fourth

(4) Mismatch error

but which is of a substantially different nature than the others and from a different point of view is not regarded as an error. In any case, these will be briefly discussed in the order listed.

Instrumentation error

In a few words, this is the error in P_{bolo} as determined by eqn. 13.2. It includes the errors in the measurement of R, I_1, I_2 and the failure of the feedback loop to maintain an exact bridge balance. Equation 13.2 may also be written

$$P_{bolo} = \frac{R}{4}(I_1 + I_2)(I_1 - I_2) \qquad (13.3)$$

Since P_{bolo} is given by the *difference* between two bias powers, the demands on the associated instrumentation can be fairly severe when these are nearly equal, as will be the case for 'small' values of microwave power. For example, if this difference is 1%, one requires I_1 and I_2 to a part in 10^4 if an accuracy of 1% is to be achieved. For this reason, the dynamic range for the bolometric technique is typically limited to a nominal 20 dB or so if the best accuracy is required.

DC-RF substitution error

The bolometric technique includes the implicit assumption that the microwave power is equal to the change in (DC) bias power or that the functional dependence of bolometer resistance on power is the same for both sources. The *current* distributions, however, will be different for the two because of skin effect and related phenomena. Thus, the distribution of heat sources throughout the bolometer will be different. In the general case, one can expect that the bolometer resistance will be a function, not only of the total power which is dissipated therein, but also of *where* (within the bolometer element) the dissipation occurs.

   This topic has been the subject of a substantial amount of theoretical study in an earlier era with the general conclusion that, for the barretter types in general use, the error could be ignored for frequencies up to 10 GHz.[28]   Moreover, the existing calibration methods are such that this error is usually evaluated along with the bolometer mount efficiency as described below.

   This phenomen is also of potential concern in the context of the dual-element bolometer mount. As shown in Fig. 13.3, a popular if not universal design in coaxial systems calls for a pair of bolometer elements which are in series at DC, but in parallel for RF. Although the frequency region where this design is useful includes that for which the criteria for the applicability of low-frequency circuit

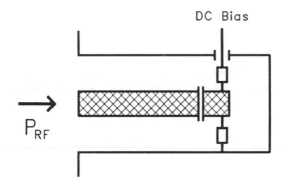

Fig. 13.3 Dual-element bolometer mount

[28]See, for example:

S. JARVIS, Jr., and J. ADAMS, 'Calculation of substitution error in barretters', *J. Res. Nat. Bur. Stand. (Eng. and Instrum.)*, 72C, pp. 127-137, April-June 1968

J. W. ADAMS, and S. JARVIS, Jr, 'Current distribution in barretters and its application to microwave power measurements,' *IEEE Trans Microwave Theory Tech.*, MTT-17, pp. 778-785, Oct. 1969

theory may be only marginally satisfied, a substantial amount of additional insight into the operation may be frequently gained by the use of these low-frequency concepts even if the results are not 'rigourous'. In general, the *series* connection for DC enforces the condition of equal *currents* in the two elements and a constant value for the *sum* of their resistances, the *parallel* connection at RF calls for equal *voltages* across them. Apart from perfect symmetry, the initial application of DC bias power will ordinarily lead to different values of individual resistance (e.g. 101Ω for one and 99Ω for the other.) In particular, it is only the *sum* of the resistance values which is maintained constant by the associated (bridge) bias circuitry. Following this, the application of RF power will also affect the resistance division between the two elements and introduce an error which is given by $(1/\gamma_2 - 1/\gamma_1)\Delta r$, where γ_1 and γ_2 are the 'ohms/milliwatt' coefficients for the individual elements and Δr is the change (if any) in resistance division which occurs (e.g. if after the application of RF the resistances are 103Ω and 97Ω, $\Delta r = 2$). It may be further noted that, because of the parallel connection, the *larger* fraction of the total RF power will be dissipated in the element of *smaller* resistance. If the temperature coefficients are positive, the resistance shift will be in the direction of equality. In the more common case of negative coefficients, the change will be in the direction of further aggravating the existing unbalance. For this reason it is usually necessary, with the dual-element thermistor mount, to limit the RF power to 25-35% of the total bias power. In general, the error from this source is dependent on the degree of care exercised by the manufacturer in matching the two bolometer elements.

Bolometer mount efficiency

Returning to Fig. 13.1, the bolometric technique, at best, measures only the power dissipated within the bolometer element. By contrast, one is usually interested in measuring the net (or possibly 'incident') power at the *input* to the bolometer mount. In general, some of this power will be dissipated in the imperfectly conducting walls and/or dielectric support structures within the mount. By definition, the ratio of the RF power dissipated in the bolometer element, to the net power input, is the bolometer mount efficiency. Although efficiencies of 95-98% are common at frequencies below 10GHz, the required correction can become much larger as one moves into the millimeter frequency region.

Effective efficiency

Unless corrected, the substitution and efficiency phenomena represent two well defined sources of systematic error. In many cases however, an example of which is given below, it is convenient to combine them into a single parameter, 'effective efficiency' η_e, which is defined by

$$\eta_e = \frac{P_{bolo}}{P_{RF}} \qquad (13.4)$$

where P_{bolo} is the power indicated by the bolometric technique and P_{RF} is the *net* power input.

The micro-calorimetric technique for measurement of effective efficiency

The micro-calorimetric technique, of evaluating effective efficiency, was first proposed by MacPherson and Kerns of the National Bureau of Standards in 1954. Today, in combination with further refinements and improvements, it has become the basis for the primary standard of microwave power measurement at many of the national laboratories. Although the technique now takes a variety of forms, one basic idea is common to all: an evaluation of η_e by calorimetric methods. The essential ideas will next be described.

The method is illustrated by Fig. 13.4, where the basic feature includes a provision for making a *simultaneous* measurement, by both bolometric and calorimetric techniques, of the *same* power. If one can assume that the calorimetric value is 'correct', the ratio between the bolometric and calorimetric determinations provides a measure of the effective efficiency. Since the power levels involved are very small, a substantial degree of sophistication is required in the associated instrumentation. In any case, the essential features of a power determination by calorimetric methods include: (1) a calorimetric body or object in which to dissipate the energy of interest; (2) provision for measuring the resulting temperature rise and; (3) a method of determining the proportionality factor between the observed temperature rise and the energy or power which is to be measured.

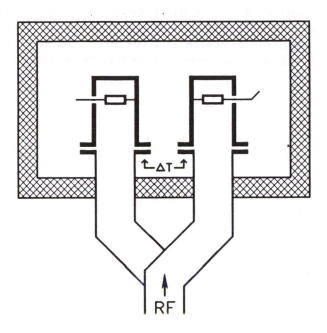

Fig. 13.4 Cross-section of microcalorimeter

In the microcalorimeter technique, the bolometer mount itself is the 'calorimetric body'. This, in turn, calls for thermal isolation or decoupling between it and its environment. Here, the biggest problem is the waveguide lead itself, which is ordinarily a good thermal conductor. This calls for the use of a short 'isolation' section of waveguide, in which the wall thickness, and thus thermal conductivity, have been greatly reduced, while the electrical conductivity is preserved. The required mechanical stability in the isolation section is then restored as necessary by the use of a suitable plastic or other thermally insulating material. The thermal isolation between the bolometer and environment may be further enhanced by use of the 'Twin-Joule' configuration, as also shown in Fig. 13.4. Here, the temperature rise is indicated by a suitable thermopile and the temperature reference point is a second or 'dummy' bolometer mount which is isolated from both sources of energy, but whose dependence on ambient temperature fluctuations is (hopefully!) the same as that for the active mount. In this way the effect of ambient temperature variations is minimised.

In use, the system is first 'calibrated' by the application of DC bias power in the bolometer element, as required to establish bridge balance. This is also the first step in a bolometric power determination. At thermal equilibrium, there will be a constant thermopile response and at this point the heat energy is being conducted to the external heat sink at the same rate as it is being supplied. Next, with the addition of microwave energy and the withdrawal of DC, (in keeping with the requirements of the bolometric technique), one observes a further increase in the thermopile response — the thermopile is responding to the energy dissipation in the mount as a whole, not just that in the bolometer element. This increase in temperature is a measure of the *difference* between the effective efficiency and unity. This, in turn, provides a correction to the bolometrically-determined power which may then be applied in its subsequent use including, for example, the calibration of other mounts or power meters.

One particular feature should be noted. As a counterpart to the DC-RF substitution error described above, one can also expect a calorimetric equivalence error or dependence of the thermopile response on the spatial distribution as well as total energy dissipation within the bolometer mount. In the present case, the energy dissipation, by which the thermopile calibration is effected, is for all practical purposes confined to the bolometer element. This is also the place where, typically, 95% or more of the RF energy is dissipated. Because the thermopile sensing is 'remote' from this location, one can expect it to be insensitive to the small differences in the RF and DC energy distributions within the bolometer element itself. Thus the response is virtually independent of the DC-RF substitution error. With respect to the remainder of the microwave energy, the calorimetric equivalence error can be minimised by constructing it of materials of high thermal conductivity as appropriate. In any case, the nonequivalence error applies only to that fraction, typically 5% or less, of the energy dissipation which occurs other than in the bolometer element. If this added power is measured to an accuracy of only 10%, the associated error has been reduced from 5% to 0.5%. In practise, outer bounds or limits for this error may be experimentally determined by the temporary addition of other localized heat sources within the bolometer mount and observing the changes in thermopile response as a function of their location.

Perhaps the most difficult problem in this context is to 'define' the interface between the calorimetric body, or bolometer mount and the heat sink. As already noted, the two are interconnected by a short section of thin-wall waveguide which, ideally, is lossless but also of high thermal resistance. It is along this short section of waveguide that the temperature difference, as measured by the thermopile, primarily exists. Although the microwave dissipation here is minimised by keeping its length short, in general there will be some contribution to the thermopile response from this source and this frequently represents the major source of error in the method.

Today these devices exist in a substantial number of national standards laboratories and in both waveguide and coaxial versions. The many refinements include automated data taking and frequency control. In some versions Peltier cooling is added as a method of heat removal. The engineering problems in implementing the technique are perhaps the easiest to handle in 10 GHz rectangular, or 'X-band' waveguide. For larger waveguides the thermal mass increases but the power levels remain constant. This aggravates the problem of ambient temperature control. At higher frequencies, or with smaller waveguides, the dissipation in the isolation section is more of a problem, and with the inevitable decrease in mount efficiency, the calorimetric equivalence error becomes more important. In coaxial versions the major problem is associated with the centre conductor, where most of the dissipation in the isolation section occurs and whose temperature profile is difficult to determine. In spite of these problems, the method typically yields accuracies of a few tenths of a percent through 10 GHz and reduced accuracies up to 100 GHz.

The 'impedance' method of measuring bolometer mount efficiency

Although current interest is focused primarily on the calorimetric technique, the existence of an alternative method of measuring mount *efficiency* (as contrasted with effective efficiency) will be briefly noted. This is the so-called 'impedance' method which is based on observing the changes in reflection coefficient, or input impedance, as a function of the bolometer operating resistance. The theoretical basis for this is briefly as follows:

The analysis begins with a 'two-port' *model* for the bolometer mount which (assuming reciprocity) may be characterised by three complex parameters. The mount *efficiency* will be a function of these parameters, which may, in turn, be determined from observations of the input impedance in combination with the corresponding values of bolometer resistance and where changes in the latter are achieved by suitable adjustments in the (DC) bias power.[29] The application requires that the equivalent circuit of the bolometer element (at microwave frequencies) is one in which the resistance is proportional to the DC value and the reactance is constant. The available evidence indicates that this is well satisfied by barretter types. Unfortunately, and for reasons not fully understood, the thermistor type bolometer element does not satisfy these criteria. In addition, the same design features which are the source of the substitution error in the dual-element bolometer mount, as described above, also make it difficult to accurately predict the (parallel) resistance to the RF source from observations of the total series resistance at DC. For these reasons, the method is limited to single element (waveguide) mounts and barretter type elements. Although this represents only a small segment of the field of potential application, the method did reach a highly developed state and found substantial use at the National Bureau of Standards prior to the advent of automation.[30] There is some reason to anticipate that it may again become a useful alternative to the calorimetric technique at millimeter wavelengths.

Reflection 'error'

For completeness, one should recognise that in general some of the incident energy will be reflected and thus not 'measured' by the technique as described. Whether or not this represents an error depends on the nature of the application.

For example, if one starts with a signal source and the objective is to determine the power which it would deliver to a matched termination, then the 'reflection' is certainly a source of error. The stated objective, however, tends to be academic in that terminations are seldom 'matched' (although this objective may be closely approximated). In a more general context, if the requirements of careful metrology are to be satisfied, it is necessary to recognise and account for the deviations from the matched condition. At this point the mount reflection no longer represents a source of error but rather an additional parameter whose value may be explicitly accounted for as described in the next chapter.

---

[29]For further details see D. M. KERNS: 'Determination of efficiency of microwave bolometer mounts from impedance data', *J. Res. Nat. Bur. Stand.* 42, pp. 579-585, June 1949

[30]G. F. ENGEN: 'A bolometer mount efficiency measurement technique', *J. Res. Nat. Bur. Stand. (Eng. and Instrum.)*, 65C, pp. 113-124, April-June 1961

Chapter 14
Power meter applications

Power measurement problems at microwave frequencies tend to fall into one of two categories: First, one may be interested in the performance characteristics of a signal source. At the lower frequencies the signal sources tend to be 'constant voltage' although, strictly speaking, this implies an infinite source of available power. In more practical terms, this means that the source impedance is very small in relation to that of the devices to which it is (usually!) connected. Thus in terms of potential power transfer, a gross 'mismatch' is present. By contrast, at microwave frequencies, the goal is usually that of obtaining the maximum power transfer and a determination of the departure from this objective frequently represents an important part of the measurement procedure. Alternatively, the problem may be that of measuring the rate of energy flow in a certain operating environment. In the metrology laboratory, the latter may take the form of making an independent determination of the power delivered to a terminating type power meter, which may then be compared with its own indication and thus effect its calibration.

As a rule, microwave power meters fall into the categories of 'terminating' and 'feed-through' types. The bolometric technique, as described in the preceding chapter, (ideally) terminates the line in its characteristic impedance. It thus absorbs the energy which it measures and falls into the first category. By contrast, the feed-through type may absorb only a fraction of the available energy and pass the remainder to a termination. The first type tends to be the basis for calibrating or assessing the operation of the second type and its applications will be described first. In general terms, the power which is indicated by a terminating type meter, when connected to a source of microwave energy, is usually used to infer the power which would be delivered to a different termination (e.g. antenna) for which the meter has been temporarily substituted. In the process, certain characteristics of the signal source (e.g. available power) may also be obtained. Since this is at least implicit, if not explicit, in the usual procedure, the use of a power meter for evaluation of the source parameters will be described first.

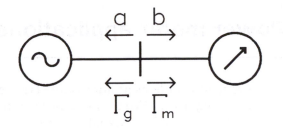

Fig. 14.1 Basic circuit

Referring to Fig. 14.1, the (net) power input, P_{gm}, to a terminating meter is given by

$$P_{gm} = |b|^2 - |a|^2 = |b|^2 \left(1 - |\Gamma_m|^2\right) \qquad (14.1)$$

where Γ_m is the reflection coefficient of the meter. After introducing the source parameters b_g and Γ_g, and by use of eqn. 3.11, this becomes

$$P_{gm} = \frac{|b_g|^2}{1 - |\Gamma_g|^2} \cdot \frac{\left(1 - |\Gamma_g|^2\right)\left(1 - |\Gamma_m|^2\right)}{|1 - \Gamma_g\Gamma_m|^2} \qquad (14.2)$$

and which by comparison with eqn. 3.20 may also be written

$$P_{gm} = P_g M_{gm} \qquad (14.3)$$

where the subscript 'm' has again been used to represent 'meter'.

As further described in Chapter 3,

$$P_g = \frac{|b_g|^2}{1 - |\Gamma_g|^2} \qquad (14.4)$$

and

$$M_{gm} = \frac{\left(1 - |\Gamma_g|^2\right)\left(1 - |\Gamma_m|^2\right)}{|1 - \Gamma_g\Gamma_m|^2} = 1 - \frac{|\Gamma_m - \Gamma_g^*|^2}{|1 - \Gamma_g\Gamma_m|^2} \qquad (14.5)$$

With reference to eqn. 14.4, P_g is a function only of the generator parameters and may be obtained from eqn. 14.3 *provided that* M_{gm}, which involves both Γ_g and Γ_m on a *complex* basis, can be determined.

To continue, the power delivered to a different termination, P_{gl}, will be given by an equation of the same form as eqn. 14.3 and in terms of the meter response may be written

$$P_{gl} = P_{gm} \frac{M_{gl}}{M_{gm}} = P_{gm} \frac{|1 - \Gamma_g \Gamma_m|^2}{|1 - \Gamma_g \Gamma_l|^2} \cdot \frac{1 - |\Gamma_l|^2}{1 - |\Gamma_m|^2} \tag{14.6}$$

which is a well known result. Here the generator parameter, P_g, no longer explicitly appears, since it has been eliminated in the ratio. Although this result, eqn. 14.6, has found extensive application in intercomparing power meters, it is important not to lose sight of the implicit assumption that P_g is a constant.

The directional coupler techniques introduced in the previous chapter provide a convenient method of monitoring this parameter

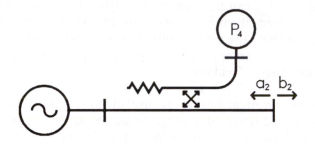

Fig. 14.2 A directional coupler may be used to monitor P_g.

(P_g). In particular, and referring to Fig. 14.2, eqn. 12.27 may be written

$$P_4 = |b|^2 |D|^2 |1 + (C/D)\Gamma|^2 \tag{14.7}$$

which may then be combined with eqn. 14.1 to yield

$$P_{gm} = \frac{P_4 \left[1 - |\Gamma_m|^2\right]}{|D|^2 |1 + (C/D)\Gamma_m|^2} \tag{14.8}$$

and which, with further manipulation, becomes

$$P_{gm} = \frac{P_4}{|D|^2 \left[1 - |\Gamma_{gs}|^2\right]} \cdot \frac{\left[1 - |\Gamma_{gs}|^2\right]\left[1 - |\Gamma_m|^2\right]}{|1 - \Gamma_{gs}\Gamma_m|^2} \tag{14.9}$$

and where Γ_{gs} has been substituted for $-C/D$. As explained in Chapter 6, this is the reflection coefficient for the 'equivalent source' which is provided by the directional coupler. Finally, eqn. 14.9 may be written

$$P_{gm} = K_A P_4 M_{gm} \qquad (14.10)$$

where

$$K_A = \left(|D|^2 \left[1 - |\Gamma_{gs}|^2 \right] \right)^{-1} \qquad (14.11)$$

As before, if M_{gm} can be determined, the parameter K_A may be obtained by use of eqn. 14.10 and the device now becomes a 'directional coupler power transfer standard' which provides the equivalent of a calibrated signal source whose available power is given by $K_A P_4$ and of source impedance Γ_{gs}. To the extent that the mismatch factor M_{gm} can be considered unity, the device may also be regarded as a feed-through power meter since, apart from the mismatch correction, the power delivered to a termination is simply $K_A P_4$. Moreover, by an appropriate choice of coupling ratio, K_A can assume a wide range of values.

For some applications it is useful to further enhance the operation by the addition of a second coupler as shown in Fig. 14.3. If tuning transformers are also added, the device becomes a tuned

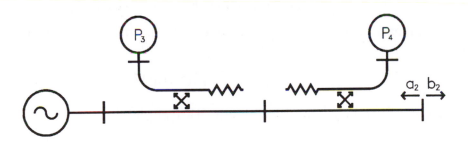

Fig. 14.3 Addition of a second coupler to enhance the operation

reflectometer as described in the previous chapter, but with the important extension that power, as well as $|\Gamma_l|$, may now be measured. In particular, the tuning procedure leads to the condition $\Gamma_{gs} = 0$ and the reflectometer provides for the measurement of $|\Gamma_l|$ so that M_{gl}, which now equals $\left(1 - |\Gamma_l|^2 \right)$, may be evaluated. Moreover, for the tuned reflectometer, one has $B = C = 0$ and with the help of eqn. 12.26, eqn. 14.1 becomes

$$P_{gl} = K_A P_4 - K_B P_3 \qquad (14.12)$$

where K_A is now just $1/|D|^2$ and $K_B = 1/|A|^2$.

A more general result

Although the foregoing is a straightforward extension of the techniques developed in the previous chapter, additional insight into the operation, where these ideal criteria are not satisfied, is provided by an alternative development.

Returning to eqn. 14.9 this may be combined with eqns. 14.5, 14.10 and 14.11 to yield

$$P_{gm} = K_A P_4 \left(1 - \frac{|\Gamma_m - \Gamma_{gs}^*|^2}{|1 - \Gamma_m \Gamma_{gs}|^2} \right) \qquad (14.13)$$

In terms of the parameters, $A \cdots D$, the *complex* reflectometer response, w, is given by[31]

$$w_m = \frac{b_3}{b_4} = \frac{A\Gamma_m + B}{C\Gamma_m + D} \qquad (14.14)$$

If this is now solved for Γ_m and recalling that $\Gamma_{gs} = -C/D$, one has

$$\frac{\Gamma_m - \Gamma_{gs}^*}{1 - \Gamma_m \Gamma_{gs}} = \frac{\left[|D|^2 - |C|^2\right] w_m - \left[BD^* - AC^*\right]}{AD - BC} \qquad (14.15)$$

As described in Chapter 9, the response of w to a sliding short will be a circle of radius R, centre R_c. In terms of the notation (including the parameter D) which is employed in this chapter, and noting that for a sliding short $r = 1$, eqns. 9.4 and 9.5 become

$$R = \frac{|AD - BC|}{|D|^2 - |C|^2} \qquad (14.16)$$

---

[31]In power measurement applications, it is usually desirable to retain the parameter, D, in the formulation. Equation 14.14 is the more general counterpart of eqn. 12.15.

and

$$R_c = \frac{BD^* - AC^*}{|D|^2 - |C|^2} \qquad (14.17)$$

With the help of these results, eqn. 14.15 becomes

$$\frac{\Gamma_m - \Gamma_{gs}^*}{1 - \Gamma_m \Gamma_{gs}} = \frac{w_m - R_c}{R} \qquad (14.18)$$

and finally, eqn. 14.13 may be written

$$P_{gm} = K_A P_4 \left(1 - \frac{|w_m - R_c|^2}{R^2} \right) \qquad (14.19)$$

while $|w_m|$ is given by

$$|w_m| = \left(\frac{P_3}{P_4} \right)_m^{\frac{1}{2}} \qquad (14.20)$$

Discussion

The result given by eqn. 14.19, which holds for a four-port network of arbitrary parameters, explicitly displays several features of interest. First, in the absence of phase response in the detectors, it is not possible to completely evaluate the expression $|w_m - R_c|$. At this point one has three choices which include: (1) accept the resulting uncertainty as an error or, as an alternative, add tuning and adjust as required such that either: (2) $R_c = 0$ or (3) $w_m = 0$. With respect to the first of these options, one has

$$|w_m - R_c|^2 = |w|^2 - wR_c^* - w^*R_c + |R_c|^2 \qquad (14.21)$$

thus, assuming a worst case phase relationship, the desired result will be in error by

$$\text{Error} = \pm \frac{2|w_m R_c|}{R^2} \qquad (14.22)$$

assuming that the contributions from both $|R_c|$ and $|w_m|$ to eqn. 14.21 have been accounted for.

If tuning is to be added, this may be conveniently done in the context of Fig. 14.3 yielding Fig. 14.4. As compared with Fig. 12.6, the tuner, T_X, is not required. The basis for this may be recognised

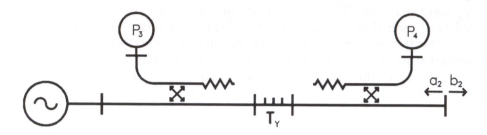

Fig. 14.4 Addition of tuning to further enhance the operation

by noting that the 'nonideal' characteristics are contained entirely in a single parameter, R_c, which can be made to vanish by use of a single tuner, T_Y. The correct adjustment can be recognised by a moving short. (Note that, if in Fig. 12.6 T_X is assumed to be lossless, the same value of (net) power obtains at either end of the tuner. Moreover, for the moving short, the reflection coefficient will be of unit magnitude at either end. On the other hand, the 'directivity' parameter (B) will certainly depend on the adjustment of T_X.)

To return to T_Y, one has two options: adjust for either $R_c = 0$ or for $w_m = 0$. The first is familiar from the tuned reflectometer operation as described in Chapter 12 and leads to the operation described by eqn. 14.12. The alternative ($w_m = 0$) simply calls for the adjustment of T_Y, with the termination of interest connected, such that P_3 vanishes. The mismatch correction may now be effected by observing the response to a sliding short,[32] which yields $|R_c|$ and R, and which may then be substituted in eqn. 14.19, where $w_m = 0$. In practical terms, the latter procedure is probably the simpler but, on the other hand, if the power to a number of different terminations is required, as in a power meter calibration environment, for example, the adjustment of T_Y must be repeated for each. By contrast, if R_c is made to vanish, this needs only to be done once.

One further observation may be made. The evaluation of the mismatch correction via eqns. 14.6 or 14.13 appears to call for a

---

[32]For a more complete discussion see Chapter 15.

determination of the arguments as well as magnitudes of Γ_g and Γ_l. Ordinarily this represents a problem because the reflectometer does not provide the phase. In the prior art, a substantial premium was placed on making Γ_g vanish such that only the magnitudes of Γ_l and Γ_m were required. The above technique, however, achieves this result without requiring an explicit determination of phase and is useful in a calibration laboratory, for example, in the context of calibrating a terminating type power meter in terms of a similar 'standard' power meter. Here, the choice of generator or signal source is usually at one's disposal.

In the more general context, however, as illustrated by Fig. 14.1, it is usually necessary to revert to eqn. 14.6 in order to determine the power delivered to a different termination (e.g. antenna) and this does involve both Γ_g and Γ_l on a *complex* basis. Due to the practical difficulties in obtaining the arguments, as contrasted with the magnitudes, there is a substantial precedent for making only a partial correction for the mismatch. This consists of evaluating only the terms $\left[1 - |\Gamma_l|^2\right]$ etc. and accepting $|1 - \Gamma_g\Gamma_l|^2$ as a 'mismatch error'. When this is done, and assuming a worst case phase, it is easily shown that

$$\frac{|1 - \Gamma_g\Gamma_l|^2}{|1 - \Gamma_g\Gamma_m|^2} \approx 1 \pm 2\left(|\Gamma_g\Gamma_l| + |\Gamma_g\Gamma_m|\right) \qquad (14.23)$$

It is also common practise, in this context of making only a *partial* mismatch correction, to characterise the source by a 'generator power' which, by definition, is simply that which it will deliver to a matched or reflectionless termination. From inspection of eqn. 14.2 this is just $|b_g|^2$. In a similar way, the counterpart of K_A, or ratio of indicated coupler sidearm power to that delivered to a matched termination on the main arm is, by definition, the 'calibration factor'.[33] Finally, the $\left[1 - |\Gamma_m|^2\right]$ correction may be combined with the bolometer mount effective efficiency to yield another 'calibration factor'. This approach to the problem is characterised by a focus on the *incident* power or $|b|^2$ as contrasted with *net* power which is given by $\left[|b|^2 - |a|^2\right]$. It also includes the *implicit assumption* that, 'the rest of the world is ideal', apart from the component of immediate interest. (e.g. if a terminating meter has been calibrated in terms of

---

[33]The reciprocal of this definition is also in use.

its 'calibration factor', it provides a direct indication of the power which a *matched* generator would deliver to a *matched* termination.)

In any case, whatever its merit or shortcomings, this technique for partial mismatch correction is deeply embedded (entrenched?) in the existing art. However, with increased accuracy demands and the introduction of techniques which yield both argument as well as magnitude, there is more incentive to make a complete mismatch correction rather than accept the interaction factor as an error. If this is to be done, there are some substantial advantages which accrue from basing the description on net power rather than incident power. These will be described in greater detail in the next chapter on *'Power eqnuation methods.'*

Extension to higher power levels

As noted in the previous chapter, the existing standards of power measurement are based primarily on the bolometric technique, in which the useful range of power levels is nominally 100 μW to 10 mW. Although in an earlier era a substantial effort was invested in the development of standards for use at higher power levels, today this is largely achieved by an extension or extrapolation from the lower power levels.

Perhaps the simplest technique for so doing is to add a two-port device to the power meter for which, ideally, $S_{11} = S_{22} = 0$, and $|S_{21}|^2$ or the attenuation is known. For example, if one starts with a *matched* bolometric power meter and adds a 20 dB attenuator, the range now becomes, 10 mW-1 W. For larger values of power (and attenuation), it may be convenient to use a directional coupler as

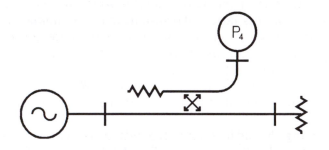

Fig. 14.5 Use of a directional coupler to provide an extension to higher power levels

shown in Fig. 14.5. Here the attenuation is provided by the coupling mechanism, and the main arm termination provides a (usually) more convenient method for disposing of the associated heat. In the general case, where the components are not ideal, it is necessary to correct for mismatch effects as described in the chapter to follow. The remainder of this chapter will consider the problem from a different perspective.

Returning to Fig. 14.5, the techniques of interest will be described by the use of a specific example. Let a nominal 40 dB coupling ratio be assumed, while the sidearm detectors and available standards are of the bolometric type. The requirement is to determine the parameter K_A. Although in principle the previously described techniques are applicable, the main guide power will be limited to a nominal 10 mW by the postulated bolometric standard. The sidearm power is thus only 1 μW, which is too small to be accurately measured. At this point the generator is connected to the coupler output port and certain components added which, as shown in Fig. 14.6, include tuners T_X, T_Y and a 20 dB coupler with associated sidearm detector. The

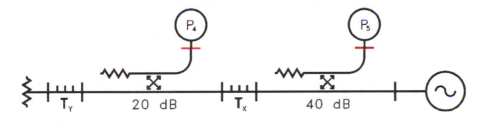

Fig. 14.6 Experimental setup for adjustment of tuner T_X

tuner, T_Y, is next adjusted for a null at P_4 and T_X is then adjusted for a null at the sidearm of the 40 dB coupler which has been denoted P_5. The sensitivity of this latter adjustment may be enhanced, either by the temporary substitution of a more sensitive detector at P_5 (e.g. a receiver if available) or, alternatively, by increasing the power level at the source. (Note that most of the power will be dissipated in the load which terminates T_Y. If the source is 10 W, only 100 mW will be dissipated in the internal termination of the 20 dB coupler.)

Following this adjustment, the system is next reconnected and further expanded as shown in Fig. 14.7, where, for improved sensitivity, a coupling ratio of 3 dB has been specified for the added coupler. As a consequence of the previous adjustment of T_X, the equivalent source impedance, which obtains at port 2, is the same for both the 20 dB coupler and the 40 dB coupler. (The 40 dB coupler

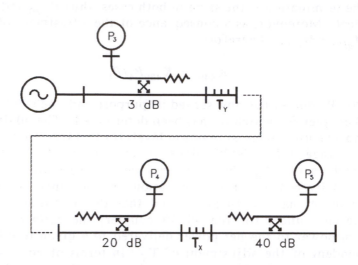

Fig. 14.7 The dynamic range may be extended to 100 W
by use of the circuit shown here

may here be regarded as an extension of the output arm of the 20 dB
coupler.) A standard terminating meter (of the bolometric type), in
combination with the techniques described above (including the
further adjustment of T_Y, but not of T_X), may now be used to
determine K_A for the 20 dB coupler. (Note that the sidearm signal
level is a nominal 100 μW, which is at the lower level of the operating
range for the assumed bolometric technique.) Moreover, the added
3 dB coupler, in combination with T_Y, allows an evaluation of the
mismatch factor.

The standard power meter is next replaced by a high power
termination of arbitrary but nominal impedance match. The 3 dB
coupler has served its role (in determining the mismatch correction)
and it is usually desirable to remove it in order to eliminate the
dissipation associated therewith. The remaining step is to increase the
input power to a nominal level of 1 W, such that the sidearm powers
become a nominal 10 mW and 100 μW for the 20 and 40 dB couplers,
respectively. If the added subscripts '20' and '40' are used to identify
the 20 and 40 dB couplers one now has

$$P_{gl20} = K_{A20}P_4M_{gl20} \qquad (14.24)$$

and

$$P_{gl40} = K_{A40}P_5M_{gl40} \qquad (14.25)$$

But the termination is the same in both cases, thus P_{gl20} and P_{gl40} are identical. Moreover, as a consequence of the adjustment of $\hat{T}_X$, one has $M_{gl20} = M_{gl40}$. Therefore

$$K_{A40} = K_{A20}P_4/P_5 \tag{14.26}$$

The 20 dB coupler has now served its purpose and one is left with the 40 dB coupler for which K_A has been determined. The 40 dB coupler may now be used to measure power levels for this specific termination in the range 1 W - 100 W, since an evaluation of the associated mismatch factor M_{gl40} has been included in the above. If the net power delivered to this termination is the only measurement objective, it may be further noted that the adjustment of T_X is unnecessary since, for the 40 dB coupler, the *ratio* of indicated sidearm power to the net power delivered to a given termination is independent of the adjustment of T_X. In terms of eqn. 14.25, this means that the $K_{A40}M_{gl40}$ product has been determined, but the individual factors have not.

Although there are obviously a lot of variants, the basic idea is one which permits an *addition* of the dynamic ranges of the respective detectors.

Chapter 15

An introduction to

Power equation methods

As noted in the previous chapter, an important part of the assessment of microwave system performance is derived from the measurement of power, and in which attenuation and impedance measurements often have a supporting role. As described therein, the role of impedance measurement is frequently that of providing a determination of the (mismatch) correction which must be applied to compensate for a failure of the different components to satisfy the 'matched' condition. These techniques, in turn, are based on microwave circuit theory, which was developed in an earlier chapter and which is based on a solution of Maxwell's equations in a *uniform* waveguide.

In addition to the 'scattering' equations, microwave circuit theory may also assume a variety of other forms, including formulations based on 'generalised' voltage and current or cascading parameters, for example. These different descriptions, however, are 'equivalent' in the sense that the information content is the same, although in different form. Moreover, except for certain singularities (the 'impedance' matrix of an 'ideal' transformer, for example), it is possible to transform from one description to another.

By contrast, there is a substantial loss or suppression of detail in going from the field to the circuit description. This loss of detail greatly restricts the domain of applicability of the circuit equations as compared with the field equations but, on the other hand, it is also this elimination of detail which leads to a greatly simplified description and puts the remaining parameters in better focus.

In addition to the field and circuit descriptions is a third description which, despite its extensive usage in microwave metrology, has perhaps not been explicitly recognised as such. Provided that the different components which comprise the system are 'matched', a particularly simple description is possible. In this description all terminations are equivalent (matched or reflectionless) while the description of a generator requires a single parameter — the power output (to a matched load). The characterisation of a two-port is also achieved by a single parameter — its attenuation if the device is passive, or gain if it is active. In addition, amplifier noise properties are completely specified in terms of noise figure or noise temperature.

Provided that the hypothesis of matched components is indeed satisfied, the utility of this scheme is self-evident. Unfortunately, although this condition is included in the design objectives of most systems, the accuracy demands in system evaluation are often such that the deviations from this idealisation cannot be ignored.

One possible response is to return to the description found in the circuit equations, which has been outlined in the preceding chapters, and this is done in many cases. An alternative response has been an attempt to 'salvage' this simple description by patching it up with mismatch corrections and related concepts. Unfortunately, however, there has been no consensus on how best to do this and the existing art is characterised by a number of competing ideas and terminologies. (A survey of the literature will disclose at least four 'kinds' of generator power, perhaps a dozen (many redundant) terms for characterising the 'loss' characteristics of a two-port and maybe a score of descriptions for amplifier noise.) Much of this appears to be rooted, at least implicitly, in the unresolved question of how to best characterise the components in a mismatched system without introducing the added detail which is a part of the circuit description.

It is the purpose of this chapter to formalise an alternative description which has actually been introduced in the earlier chapters, although not specifically identified as such. This description is based on *power equations* and explicitly accounts for the mismatch effects and avoids much of the complexity of the circuit equations. It includes, as a special case, the simple description of a matched system as outlined above. In the power equation description the *net power*, in the waveguide of interest, is the fundamental and *real* parameter. By contrast, the circuit approach uses two *complex* parameters (e.g. the forward and reverse wave amplitudes, or alternatively, v, i) to describe the behaviour at a given terminal surface. It will thus be immediately recognised that the power equation description includes a further suppression of detail, but as a consequence of this the remaining parameters are placed in better focus. Although the problems to which this description is applicable are few in number, their practical importance is great. The power equation description provides:

(1)   A systematic basis for the choice of parameters to describe the behaviour of mismatched systems

(2)   Added insight into 'mismatch' and related measurement problems

(3)   Simplified procedures for the evaluation of mis-match corrections

(4)   *Virtual elimination of the 'uniform waveguide' requirement and reduced sensitivity to certain types of connector problems*

Theoretical background

As shown in Fig. 15.1, it is convenient to postulate a 'circuit' which is divided into three regions. Starting from the left, these include one which contains an enforced current (or source) **J**, a loss free region, and a 'load', where both the first and last of these include energy absorbing dielectric materials. It is also assumed that the

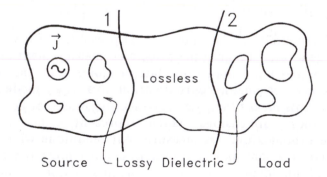

Fig. 15.1 *'Circuit' for discussion of power equation methods*

complete system is contained within a perfectly conducting envelope of arbitrary geometry. In general, **J** will be the source of an electromagnetic field throughout the entire structure and thus deliver energy to the lossy dielectric materials in the third region. In one form (mode) or another the power, which will be denoted by P_{gl}, *must* be propagated through the postulated lossless region. Moreover, conservation of energy requires that the power crossing 'terminal surface' 2 in Fig. 15.1 is the same as that at terminal surface 1. The parameter P_{gl} is thus *terminal invariant*. That is, its value is the same for all choices of terminal surface within the lossless region.

On physical grounds, one can, in general, expect that the power which is delivered to the termination will depend on the distribution and choice of lossy materials and, moreover, that there will be some configuration for which this will be a maximum. By definition, this parameter is the available power, P_g. As before, its value will be the same irrespective of whether the lossless region is considered to be part of the source or part of the termination. Thus P_g is also terminal invariant. Finally, the parameter M_{gl} is, by definition, the ratio of the actual to available power (P_{gl}/P_g). Thus it also shares the terminal invariant character of P_{gl} and P_g. In particular, although P_g, P_{gl} and M_{gl} continue to be well defined, the basis for microwave circuit theory has disappeared. This means that the concepts of wave amplitude, reflection coefficient, VSWR, directivity etc. have no meaning in the power equation context. The basic tools at one's disposal are limited to a terminating power meter (of undefined input 'impedance') and a sliding short or variable 'reactance'. Although the sliding short is usually realised in the context of uniform waveguide, the uniformity is only an operational convenience and not a fundamental requirement. The essential element, in any case, is to provide a collection of terminations for which $P_{gl} = 0$.

Unfortunately it is not possible, at present, to fully utilise this increased generality because the associated measurement techniques, which have been developed to date, require that the associated waveguide is a 'reasonable approximation' to a single mode structure. (In practise this means that the cross-section must not be allowed to approach those dimensions which represent either cut-off or multimode propagation.) A measurement technique which satisfies the terminal invariant (power equation) criteria must also be valid in the limit of an ideal waveguide, For this reason, it has proven convenient to utilise the existing microwave circuit description in the development of measurement techniques. However, in this description, as will be illustrated below, the reference to the wave amplitudes, scattering parameters and other circuit concepts must ultimately be eliminated. Although these techniques do not fully exploit the potential generality of the method, the arguments given above do provide the basis for relaxing the tolerances on waveguide cross-section, connector properties etc.

In order to illustrate the terminal invariant feature, in the circuit theory environment, it is convenient to begin with Fig. 15.2.

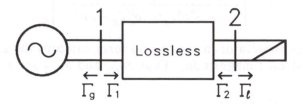

Fig. 15.2 Illustration of 'terminal invariance'

Here a generator, an arbitrary but lossless two-port and termination are connected in series. Terminal planes 1 and 2, which define the two-port, have also been specified. It will be shown that the mismatch term M_{gl1}, evaluated at terminal 1, is equal to M_{gl2}, evaluated at terminal 2.

At terminal 1 the reflection coefficient, Γ_1, is given by

$$\Gamma_1 = \frac{a\Gamma_l + b}{c\Gamma_l + 1} \tag{15.1}$$

where a, b and c are the two-port parameters. For a *lossless* two-port, however, it will be recalled (eqns. 4.31, 4.33) that $b = ac^*$ and $|a| = 1$, thus eqn. 15.1 may be written

$$\Gamma_1 = \frac{e^{j\theta}(\Gamma_l + c^*)}{c\Gamma_l + 1} \tag{15.2}$$

where $\theta = \text{Arg}(a)$. Using Γ_1 in place of Γ_l in eqn. 3.19 (or in place of Γ_m in eqn. 14.5) gives

$$M_{gl1} = 1 - \left| \frac{e^{j\theta}(\Gamma_l + c^*) - (1 + c\Gamma_l)\Gamma_g^*}{1 + c\Gamma_l - e^{j\theta}(\Gamma_l + c^*)\Gamma_g} \right|^2 \tag{15.3}$$

At terminal 2, and looking back towards the generator, the source impedance is given by

$$\Gamma_2 = \frac{e^{j\theta}\Gamma_g - c}{-e^{j\theta}c^*\Gamma_g + 1} \tag{15.4}$$

and substituting this for Γ_g in eqn. 3.19 gives

$$M_{gl2} = 1 - \left| \frac{\Gamma_l(1 - e^{-j\theta}c\Gamma_g^*) - e^{-j\theta}\Gamma_g^* + c^*}{1 - e^{j\theta}c^*\Gamma_g - (e^{j\theta}\Gamma_g - c)\Gamma_l} \right|^2 \tag{15.5}$$

Although the numerator in the second term contains the added factor $e^{-j\theta}$, this is of unit magnitude, Thus M_{gl1} and M_{gl2} are equal.

Equivalent circuit

The application of this result in the measurement environment is facilitated by the formal introduction of an *equivalent circuit*[34]

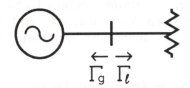

Fig. 15.3 Generator and termination

element in the form of a *lossless* two port (as utilised in Fig. 15.2, for example) but possibly of variable parameters. Starting with the model shown in Fig. 15.3, which is simply a matched generator and matched termination, the generator is completely characterised by the power output. This system is next generalised to account for mismatch effects by the introduction of lossless elements (tuners), as shown in Fig. 15.4. The generator now includes two components: a matched source (as in Fig. 15.3) and a lossless two–port or tuner, the elements of which have been chosen to yield the required value of generator

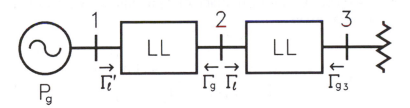

Fig. 15.4 Addition of lossless (LL) elements to Fig. 15.3

---

[34]The use of equivalent circuits is, of course, a well established practice. Among the more common examples is the use of a 'T' or 'Pi' to represent an arbitrary two-port network.

impedance. A similar observation applies the load.[35] It is noted, however, that the parameters of these lossless elements are not uniquely determined since this would require three quantities and the complex generator or load impedance determines only two. In particular, it would be possible to insert an arbitrary length of line (phase shift) between the tuner and generator (or load) without affecting the model. In addition to the physically determined (or actual) interface between generator and load (which has been designated terminal surface 2 in Fig. 15.4), it is also desirable to introduce terminal surfaces between the lossless elements and the generator (or load) that comprise the equivalent circuit. These have been designated 1 and 3 and are useful conceptually even though they have no real existence. Finally, to is convenient to characterise the generator in terms of the power that it would deliver to a matched load connected at terminal 1. Although this is not possible (except conceptually), this same power is 'available' at terminal 2 and is obtained if the termination has been chosen such that the reflection coefficient Γ_l', looking to the right at terminal 1, vanishes. This quantity, of course, is the *available power* and is the maximum power the generator will deliver as a function of load impedance. It is worth noting that the available power is dependent only on that portion of the system to the left of terminal 1, whereas the generator impedance is determined by that portion between terminals 1 and 2.

Returning to Fig. 15.4, the parameters of the lossless element included between terminals 2 and 3 must be chosen such that the reflection presented at terminal 2 is just Γ_l when its output is terminated by a match. Thus if a, b and c are redefined as the parameters which describe this two-port and noting that by comparison with eqn. 15.1, Γ_1 is now Γ_l while Γ_l has become zero, one has

$$\Gamma_l = b = ac^* \tag{15.6}$$

Moreover, $a = e^{j\theta}$ since by hypothesis the two-port is lossless. Next, looking to the left at terminal 3 one has for the reflection coefficient of the equivalent generator

$$\Gamma_{g3} = \frac{a\Gamma_g - c}{-b\Gamma_g + 1} \tag{15.7}$$

---

[35]That such a realisation is possible is presumable self-evident since a lossless transformer in conjunction with a matched load (or any load that is not completely reactive) can, in principle, be adjusted for any desired impedance.

but which, in combination with eqn. 15.6, becomes

$$\Gamma_{g3} = \frac{e^{j\theta}(\Gamma_g - \Gamma_l^*)}{1 - \Gamma_g\Gamma_l} \qquad (15.8)$$

and finally at terminal 3, since the termination is 'matched' one can write

$$P_{gl} = P_g\left[1 - |\Gamma_{g3}|^2\right] \qquad (15.9)$$

which may also be written

$$P_{gl} = P_g\left\{1 - \left|\frac{\Gamma_l - \Gamma_g^*}{1 - \Gamma_g\Gamma_l}\right|^2\right\} \qquad (15.10)$$

and which is the expression one would obtain at terminal 2. Thus, the terminal invariant character of M_{gl} has again been demonstrated. The objective of this exercise, however, has been to illustrate another feature, namely that the power equation formulation is one where the lossless two-ports, which were introduced in Fig. 15.4 to model the mismatch, have been combined into a single real parameter, $|\Gamma_{g3}|$, which contains the mismatch information in a more concise form and which then becomes the measurement objective. Moreover, with an appropriate change in notation, the circuit of Fig. 15.4 may be replaced by that of Fig. 15.2. This is a specific example of the loss of detail associated with the power equation description. At this point it should be observed that if one is interested in predicting the generator performance with arbitrary, or as yet unspecified terminations, one needs *both* P_g and Γ_g. Alternatively, if one is given the termination and wishes to describe its operation with arbitrary generators, the important parameter is Γ_l. However, once the generator and termination have been chosen, the only parameters that need be measured, when the sole objective is a determination of the net power (P_{gl}), are P_g and M_{gl}.

Application to power transfer

As a additional illustration, it is convenient to return to Chapter 14 and eqn. 14.19 which has already been formulated in accordance with the power equation criteria. In the present context this becomes,

$$P_{gl} = K_A P_4 \left[1 - \frac{|w - R_c|^2}{R^2} \right]$$ (15.11)

Here the net power, to a termination of arbitrary impedance, is given in terms of the system response P_4 and the complex w, while the system parameters are the real K_A, R and the complex R_c. As given by eqns. 14.16 and 14.17, in the limit of ideal waveguide these may be expressed in terms of the system parameters $(A \cdots D)$ and which in turn are functions of the scattering parameters. As an alternative, however, and this is the key point, these parameters can also be defined (and experimentally evaluated) in a manner which is independent of circuit considerations.

To be specific, the relationship between R_c, R and the experimentally observed w is as shown in Fig. 15.5. Here the angle between R and R_c is a function of the short position. Recalling that

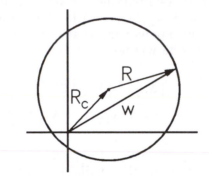

Fig. 15.5 Relationship between R, R$_c$ and w

three points determine a circle, if w is obtained on a *complex* basis, the response to three or more short positions will suffice to determine both R and the complex value of R_c. Then, given the indication (P_{gm}) of a terminating power meter (of arbitrary 'impedance'), its response can be used to evaluate K_A in eqn. 15.11 in terms of the response P_4 and w. On the other hand, if the measuring system provides only the magnitude of w, it is generally necessary to employ tuning methods, as described in Chapters 12 and 14, for example, such that either w (for the termination of interest) or R_c vanishes.

It is important to recognise that, although the discussion in Chapter 14 has focused on the use of tuning procedures, and which avoid the phase detection requirement, the need for tuning may be eliminated by providing the phase. With the growing availability of phase response, this is becoming an increasingly viable alternative.

Extension to two-port measurements

The extension of these techniques to two-port measurements begins with an arbitrary two-port, possibly nonreciprocal, which has been inserted between the generator and termination of Fig. 15.3 and

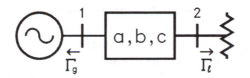

Fig. 15.6 Extension to two-port devices

where the terminals have been labelled as shown in Fig. 15.6. In the power equation context, two parameters are of primary interest. By definition the *efficiency*, η_{al}, is the ratio of the (net) power output, at terminal 2, to the (net) power input at terminal 1. By comparison with eqn. 5.13 this may be written

$$\eta_{al} = \frac{|S_{21}|^2 \left[1 - |\Gamma_l|^2\right]}{|1 - S_{22}\Gamma_l|^2 - |(S_{12}S_{21} - S_{11}S_{22})\Gamma_l + S_{11}|^2}$$

(15.12)

The other parameter of interest is the *available power ratio*, q_{ga}, or more specifically the ratio of the available power at the output port (2), to the available power from the generator. From eqn. 5.17, this is given by

$$q_{ga} = \frac{|S_{21}|^2 \left[1 - |\Gamma_g|^2\right]}{|1 - S_{11}\Gamma_g|^2 - |(S_{12}S_{21} - S_{11}S_{22})\Gamma_g + S_{22}|^2}$$

(15.13)

In eqns. 15.12 and 15.13 the results of interest depend on Γ_l and Γ_g respectively. However, there will be some choice of Γ_l and Γ_g, (which depends *only* on the two-port parameters) for which η_{al} and q_{ga} are a maximum. These may be denoted by $\Gamma_{l(max)}$, $\Gamma_{g(max)}$ and may also be considered 'reference impedances' which are physically determined by the properties of the two-port.

A further development of the two-port is given in Fig. 15.7. Here, one has a cascade of three 'two-ports' (L, M, N) the scattering

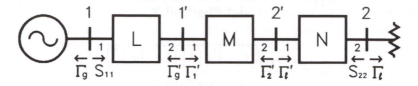

Fig. 15.7 Two-port model

coefficients of which will be denoted by $l_{11}, \cdots n_{22}$, and where the L-M and M-N interfaces are denoted as terminals 1' and 2', respectively. The first and third (L, N) of these are *lossless*; the centre element (M) is *matched*. Thus $m_{11} = m_{22} = 0$. Moreover, it is possible to choose terminals 1' and 2' in such a way that the lossless two-ports are characterised by a single (complex) parameter for each. Although a variety of decompositions are possible, a convenient one is to let

$$l_{11} = d \tag{15.14}$$

$$l_{22} = -d^* \tag{15.15}$$

$$l_{12} = l_{21} = \sqrt{1 - |d|^2} \tag{15.16}$$

$$n_{11} = e \tag{15.17}$$

$$n_{22} = -e^* \tag{15.18}$$

$$n_{12} = n_{21} = \sqrt{1 - |e|^2} \tag{15.19}$$

The scattering parameters of this cascade combination are given by

$$S_{11} = \frac{d + m_{12}m_{21}e}{1 + m_{12}m_{21}d^*e} \tag{15.20}$$

$$S_{22} = -\frac{e^* + m_{12}m_{21}d^*}{1 + m_{12}m_{21}d^*e} \tag{15.21}$$

$$S_{12} = \frac{m_{12}\sqrt{\left[1 - |d|^2\right]\left[1 - |e|^2\right]}}{1 + m_{12}m_{21}d^*e} \tag{15.22}$$

$$S_{21} = \frac{m_{21}\sqrt{\left[1 - |d|^2\right]\left[1 - |e|^2\right]}}{1 + m_{12}m_{21}d^*e} \tag{15.23}$$

and the requirement that these are equivalent to those of the original two-port represents an implicit requirement on the choice of d, e, m_{12} and m_{21}.

In order to make this explicit, and demonstrate that this is always possible, it is necessary to solve eqns. 15.20 $\cdots$ 15.23 for d, e, m_{12} and m_{21}. First, to simplify the notation, and in keeping with eqns. 4.12 $\cdots$ 4.14, let

$$a = S_{12}S_{21} - S_{11}S_{22}, \quad b = S_{11}, \quad c = -S_{22} \tag{15.24}$$

and

$$H = \frac{1 + |a|^2 - |b|^2 - |c|^2}{2|a - bc|} \tag{15.25}$$

Next, it may be confirmed by substitution that the solution of eqns. 15.20 $\cdots$ 15.23, for $|m_{12}m_{21}|$, d and e is given by

$$\eta_r = |m_{12}m_{21}| = H - \sqrt{H^2 - 1} \tag{15.26}$$

$$d = \frac{b - ac^*}{1 - |c|^2 - \eta_r|a - bc|} \tag{15.27}$$

$$e = \frac{c^* - a^*b}{1 - |b|^2 - \eta_r|a - bc|} \tag{15.28}$$

Although these results, (eqns. 15.26 $\cdots$ 15.28) are sufficient for what follows, it is possible to complete the solution by the additional use of eqns. 15.22 and 15.23. In particular,

$$\frac{m_{12}}{m_{21}} = \frac{S_{12}}{S_{21}} \tag{15.29}$$

and

$$S_{12}S_{21}\left[1 + m_{12}m_{21}d^*e\right]^2 = m_{12}m_{21}\left[1 - |d|^2\right]\left[1 - |e|^2\right] \quad (15.30)$$

After substituting the values given by eqns. 15.27 and 15.28 for d and e, this last expression is quadratic in, and may be solved for, the product $m_{12}m_{21}$. This result may then be combined with eqn. 15.29 to obtain m_{12} and m_{21}. As noted, there is no need for these results in what follows, but this does demonstrate that it is always possible to model the two-port in the indicated manner.

Returning to Fig. 15.7, because 'L' and 'N' are lossless, the efficiencies and available power ratios are *invariant* to a shift in the terminal planes from 2 to 2' and from 1 to 1'. Thus it is possible to write eqns. 15.12 and 15.13 in a simplified and more convenient form

$$\eta_{al} = \frac{|m_{21}|^2\left[1 - |\Gamma_l'|^2\right]}{1 - |m_{12}m_{21}\Gamma_l'|^2} \quad (15.31)$$

$$q_{ga} = \frac{|m_{21}|^2\left[1 - |\Gamma_g'|^2\right]}{1 - |m_{12}m_{21}\Gamma_g'|^2} \quad (15.32)$$

where Γ_l' and Γ_g' are the are the reflection coefficients of the termination and source at terminals 2' and 1'.

By hypothesis, the two-port is *passive* which requires $|m_{12}m_{21}| < 1$. (Moreover, for the *active* case (amplifier), which will be considered in Chapter 21, this is ordinarily required to avoid instability problems.) Then, by inspection of eqns. 15.31 and 15.32,

$$\eta_{max} = q_{max} = |m_{21}|^2 \quad (15.33)$$

and these are realised, in turn, when $\Gamma_l' = 0$ or $\Gamma_g' = 0$ and where

$$\Gamma_l' = \frac{\Gamma_l + e}{1 + e^*\Gamma_l} \quad (15.34)$$

while Γ_g' is given by

$$\Gamma_g' = \frac{\Gamma_g - d^*}{1 - d\Gamma_g} \quad (15.35)$$

From these results it easily follows

$$\Gamma_{l(max)} = -e \tag{15.36}$$

and

$$\Gamma_{g(max)} = d^* \tag{15.37}$$

where d and e are given by eqns. 15.27 and 15.28.

It is now possible to write

$$\eta_{al} = \eta_{max} N_{al} \tag{15.38}$$

and

$$q_{ga} = q_{max} N_{ga} \tag{15.39}$$

where, by use of eqn. 15.26

$$N_{al} = \frac{1 - |\Gamma_l'|^2}{1 - \eta_r^2 |\Gamma_l'|^2} \tag{15.40}$$

and

$$N_{ga} = \frac{1 - |\Gamma_g'|^2}{1 - \eta_r^2 |\Gamma_g'|^2} \tag{15.41}$$

Returning to Fig. 15.7, one can write for the power (P_{gl}) delivered to the termination

$$P_{gl} = P_g M_{ga} \eta_{al} = P_g M_{ga} \eta_{max} N_{al} \tag{15.42}$$

where M_{ga} is the mismatch factor, evaluated at terminal 1, between the generator and 'two-port-load' combination. The validity of this result is easily recognised since the product $P_g M_{ga}$ gives the power delivered to the two-port, whereas η_{al} is, by definition, the ratio of this power to P_{gl}. In this description, the two-port is 'combined' with the termination to yield a new 'termination' such that eqn. 3.20 applies at terminal 1 in Fig. 15.7.

Alternatively, it is possible to write

$$P_{gl} = P_g q_{ga} M_{al} = P_g N_{ga} q_{max} M_{al} \tag{15.43}$$

where P_{gl} and P_g are defined as before, $P_g q_{ga}$ is the available power from the generator-two-port combination (at terminal 2) and M_{al} is the mismatch factor between this 'modified' generator and the termination. It should be noted that eqn. 3.20 has been applied at terminal 2 rather than 1. Moreover, the parameter q_{ga} is the ratio between the *available* power at terminal 2 and the *available* power at terminal 1. By contrast, η_{al} is a ratio between *net* powers. In spite of this distinction, when reciprocity is satisfied, q_{ga} and η_{al} are related in a rather simple manner. In Fig. 15.7, η_{al} is defined in conjunction with an assumed power flow from terminal 1 to terminal 2 and is a function of the termination, but not of the source impedance. If in Fig. 15.7 one assumes that the signal source has been 'turned off', such that it becomes a passive load of reflection Γ_g, and excitation is applied via port 2, the 'efficiency' for power flow in this reverse direction (assuming reciprocity) is equal to q_{ga} and is a function of Γ_g.

The parameters N_{ga} and N_{al} may be interpreted as a measure of the extent to which q_{ga} and n_{al} differ from their maximum possible values and may thus be regarded as an alternative type of mismatch factor. Loosely speaking, their values are determined by the impedance conditions found at terminals 1 and 2, respectively. The same, however, could also be said of M_{ga} and M_{al}. The distinction lies in that N_{al} is a function *only* of the parameters of the two-port and termination while M_{al} *also* depends on the generator parameters. A similar statement could be made in reference to N_{ga} and M_{ga}. In summary, the 'M' type of mismatch depends, to some extent at least, on all of the components found in the system. As will become apparent in what follows, however, the difference between it and the 'N' type may be insignificant where attenuations of 20 dB or more are involved. With reference to the terminal invariant properties of η_{max} and q_{max} these are determined entirely by the properties of (two-port) 'M' and thus independent of the possible addition of lossless two-ports at the input terminals. (Note that these would affect only the parameters of 'L' or 'N'.) With reference to the values of Γ_l or Γ_g for which η_{max} and q_{max}, are realised, these are determined by 'N' and 'L', respectively, and, of course, subject to modification by the addition of lossless elements. However, once a value of Γ_l or Γ_g has been physically chosen, the determination of N_{al} or N_{ga} is independent of the choice of reference plane (e.g. 2 vs. 2' or 1 vs. 1') at which the evaluation is made.

Two-port measurement methods

Provided that reciprocity obtains, which is often the case, one has from eqns. 15.26 and 15.33

$$\eta_r = \eta_{max} = q_{max} \tag{15.44}$$

In addition, if the value of η_r is not too small, (which in combination with eqn. 15.24 represents the case of primary interest) the measurement of η_{al}, η_r, q_{ga}, N_{ga} and N_{al} may be effected by the use of a pair of directional couplers or 'reflectometer', as shown in Fig. 15.8, including the optional tuner, T_Y, which is useful in the absence of phase response. In keeping with the existing terminology, the *complex* system response will be denoted by w and the terms R_1, R_{c1} and R_2, R_{c2} will represent the parameters of the circular locus obtained in response to a sliding short on ports 1 and 2, respectively, as defined in Fig. 15.8.

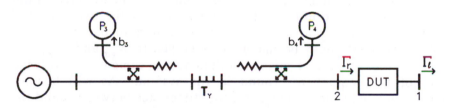

Fig. 15.8 *Measurement system for* η_a, η_{al}, q_{ga}, N_{al} *and* N_{ga}

In terms of Γ_r, which obtains at terminal 2, the response w is given by

$$w = \frac{A\Gamma_r + B}{C\Gamma_r + D} \tag{15.45}$$

where $A \cdots D$ are the network parameters and Γ_r is related to Γ_l by

$$\Gamma_r = \frac{a\Gamma_l + b}{c\Gamma_l + 1} \tag{15.46}$$

and where a, b, c are the parameters of the two-port which is defined by terminals 1 and 2 in Fig. 15.8 and whose efficiency is required for the (arbitrary) termination Γ_l.

From eqn. 14.16, R_2 will be given by

$$R_2 = \frac{|AD - BC|}{|D|^2 - |C|^2} \tag{15.47}$$

The substitution of eqn. 15.46 in eqn. 15.45 yields

$$w = \frac{(Aa + Bc)\Gamma_l + (Ab + B)}{(Ca + Dc)\Gamma_l + (Cb + D)} \tag{15.48}$$

which is in the same form as eqn. 15.45, so that

$$R_1 = \frac{|(Aa + Bc)(Cb + D) - (Ab + B)(Ca + Dc)|}{|Cb + D|^2 - |Ca + Dc|^2}$$

$$= \frac{|AD - BC| \, |a - bc|}{|Cb + D|^2 - |Ca + Dc|^2} \tag{15.49}$$

The ratio of eqn. 15.49 to eqn. 15.47 now yields

$$\frac{R_1}{R_2} = \frac{|a - bc| \left(1 - |\Gamma_g|^2\right)}{|1 - b\Gamma_g|^2 - |a\Gamma_g - c|^2} \tag{15.50}$$

where Γ_g has been substituted for $-C/D$. After another change in notation, as given by eqn. 15.24, and since reciprocity has been assumed, this result is now equal to eqn. 15.13 which is q_{ga}. Thus

$$q_{ga} = R_1/R_2 \tag{15.51}$$

If, in accordance with eqn. 14.10, the available power at terminal 2 is $K_A P_4$, that at terminal 1 is $K_A P_4 R_1/R_2$. This result is of interest in that q_{ga} is one parameter which requires neither phase information nor tuning to effect its measurement.

An equation for η_{al}, in terms of w, R_1, R_{c1}, R_2 and R_{c2}, may now be easily obtained. Since P_{gl} and P_g represent the same values in eqns. 15.42 and 15.43, it immediately follows

$$M_{ga}\eta_{al} = q_{ga}M_{al} \tag{15.52}$$

The terms M_{al} and M_{ga} are evaluated at terminals 1 and 2, respectively, in Fig. 15.8. By comparison with the last factor in eqn. 14.19, they may be expressed in terms of w, R_1, R_{c1}, R_2 and R_{c2}. Moreover, q_{ga} is given by eqn. 15.51 so that after solving for η_{al}, eqn. 15.52 may be written

$$\eta_{al} = \frac{R_1}{R_2} \cdot \frac{\left[1 - \dfrac{|w - R_{c1}|^2}{R_1{}^2}\right]}{\left[1 - \dfrac{|w - R_{c2}|^2}{R_2{}^2}\right]} \tag{15.53}$$

Provided that reciprocity obtains, the value of η_r may now be obtained from eqn. 15.53 as follows. As implicit in eqns. 15.45 and 15.46, there will be a unique relationship between w and the termination. For some value of load impedance, η_{al} will be a maximum (η_r) and this will be reflected in a corresponding value for w. Thus, conceptually, the determination of η_r is a straightforward exercise in differential calculus. The effort involved, however, may be substantial unless certain 'short-cuts' are observed.

As it stands, eqn. 15.53 is not an analytic function of w because of the absolute value operation. However, if x and y represent the real and imaginary parts of w, it is possible to consider w and w^* as two different functions of the independent variables x, y. Equation 15.53 can then be written as an analytic function of w and w^*,

$$\eta_{al} = \frac{R_2\left[R_1{}^2 - \left[(w - R_{c1})\right]\left[w^* - R_{c1}^*\right]\right]}{R_1\left[R_2{}^2 - \left[(w - R_{c1})\right]\left[w^* - R_{c1}^*\right]\right]} \tag{15.54}$$

As shown in the Appendix, a necessary and sufficient condition that η_{al} have an extremum is that

$$\frac{\partial \eta_{al}}{\partial w} = 0 \tag{15.55}$$

and in which w^* is assumed to be a constant. In combination with eqn. 15.54 this leads to

$$\frac{R_1{}^2 - |w - R_{c1}|^2}{R_2{}^2 - |w - R_{c2}|^2} = \frac{w - R_{c1}}{w - R_{c2}} \tag{15.56}$$

as the condition on w for which η_{al} is a maximum (η_r).

It is thus permissible to substitute η_r for η_{al} in eqn. 15.54 provided that w satisfies eqn. 15.56. The remaining step is to eliminate w between eqns. 15.54 and 15.56. In theory, eqn. 15.56 would be solved for w and substituted in eqn. 15.54, but this approach is tedious. Comparison of eqns. 15.54 and 15.56 shows that

$$\eta_r = \frac{R_2\left[w - R_{c1}\right]}{R_1\left[w - R_{c2}\right]} \tag{15.57}$$

After cross-multiplication and cancellation, eqn. 15.56 may be written

$$\frac{w - R_{c1}}{w - R_{c2}} = \frac{R_1^{\,2} + \left[w - R_{c1}\right]\left[R_{c1}^* - R_{c2}^*\right]}{R_2^{\,2}} \tag{15.58}$$

and this may be combined with eqn. 15.57 to obtain

$$\eta_r = \frac{R_1^{\,2} + \left[w - R_{c1}\right]\left[R_{c1}^* - R_{c2}^*\right]}{R_1 R_2} \tag{15.59}$$

At this point, eqn. 15.57 is solved for w and substituted into eqn. 15.59. This yields

$$\eta_r = \frac{G}{1 - \sqrt{1 - G^2}} \tag{15.60}$$

where

$$G = \frac{2R_1 R_2}{R_1^{\,2} + R_2^{\,2} - \left|R_{c2} - R_{c1}\right|^2} \tag{15.61}$$

Provided that the detection system provides the complex w, the measurement of R_i and R_{ci} may be effected as previously described in the context of Fig. 15.5. If only the magnitude $\left|w\right|$ is available, inspection of Fig. 15.5 yields

$$R + \left|R_c\right| = \left|w\right|_{max} \tag{15.62}$$

and

$$R - \left|R_c\right| = \pm\left|w\right|_{min} \tag{15.63}$$

As shown in Fig. 15.5, the upper sign is used if $R > \left|R_c\right|$. From eqns. 14.16 and 14.17, this will be the case if

$$\left|AD - BC\right| > \left|BD^* - AC^*\right| \tag{15.64}$$

Then, squaring both sides and cancellation yields

$$|AD|^2 + |BC|^2 > |BD|^2 + |AC|^2 \qquad (15.65)$$

which may be simplified to obtain

$$\left[|A|^2 - |B|^2\right]\left[|D|^2 - |C|^2\right] > 0 \qquad (15.66)$$

Thus $R > |R_c|$ provided that $|A| > |B|$ and $|D| > |C|$. In terms of eqns. 7.3 and 7.4 this means that b_3 must be more tightly coupled to a_2 than to b_2, while the converse must be true for b_4. These criteria are usually well satisfied.

In the absence of the phase of w, one can determine η_{al} for a termination connected to port 1 in Fig. 15.8, by first using the tuner, T_Y, to make P_3 vanish. Thus w may be set to zero in eqn. 15.53 and only the magnitudes are involved. In eqn. 15.61, R_{c2} (or R_{c1}) may be made to vanish by adjusting T_Y such that the ratio P_3/P_4 is constant with the moving short connected to port 2 (or port 1) in Fig. 15.8. With reference to eqn. 15.51, neither phase response nor tuning is required, however the q_{ga} thus obtained is for the 'Γ_g' associated with the output coupler in Fig. 15.8. If desired this may be modified, as described in Chapter 6, by the addition of a tuner at port 2.

Determination of $\Gamma_{g(max)}$ and $\Gamma_{l(max)}$

To complete this discussion, and in anticipation of the applications to be described in Chapter 21, a method of determining $\Gamma_{g(max)}$ and $\Gamma_{l(max)}$ will be given.[36] Referring to Fig. 15.9, let the 'DUT' be modeled as per Fig. 15.7 and let the termination Γ_l be

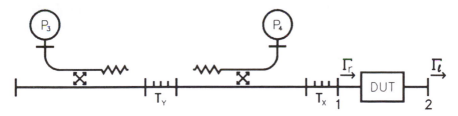

Fig. 15.9 Measurement system for $\Gamma_{g(max)}$ *and* $\Gamma_{l(max)}$

---

[36] Strictly speaking, the measurement of reflection coefficient has no meaning in the *power equation* context. On the other hand, these results are of interest in bridging the gap between these methods and microwave circuit theory.

replaced by a moving short. Since 'N' is lossless and 'M' is 'matched, at terminals 2 and 2' one has $|\Gamma_l| = |\Gamma_l'| = 1$, while at terminal 1', $|\Gamma_1'| = |m_{12}m_{21}|$. Assuming that Γ_r is obtained at terminal 1, the radius and centre of the corresponding circular locus is given by eqns. 9.4 and 9.5 where $r = |m_{12}m_{21}|$. It is possible to eliminate r between these equations by first solving eqn. 9.5 for r^2 and substituting in the denominator (only) of eqn. 9.4 The resulting expression is then solved for r, squared and substituted back into eqn. 9.5. This leads to

$$(b^* - R_c^*)(1 - R_c c/a) = R^2 c/a \tag{15.67}$$

where a, b and c are the parameters of two-port 'L'. In terms of the parameters introduced by eqns. 15.14 $\cdots$ 15.16, one has $a = 1$, $b = d$ and $c = d^*$. After making these substitutions, eqn. 15.67 becomes

$$R_c d^{*2} - \left[1 - R^2 + |R_c|^2\right]d^* + R_c^* = 0 \tag{15.68}$$

After solving for d^*, this may be combined with eqn. 15.37 to yield

$$\Gamma_{g(max)} = d^* = \frac{2R_c^*}{1 - R^2 + |R_c|^2 + \sqrt{\left[1 - R^2 + |R_c|^2\right]^2 - 4|R_c|^2}} \tag{15.69}$$

This argument may be repeated with the moving short connected to terminal 1 and leads to an identical expression for $\Gamma_{l(max)}$ in terms of R and R_c, but which now represent the parameters of the circle observed at terminal 2. If only the magnitude of $\Gamma_{g(max)}$ is required, this may be obtained from the maximum and minimum readings of a tuned reflectometer in response to a moving short, as described in Chapter 12. If the phase of $\Gamma_{g(max)}$ is also required, the preferred technique is probably one of the methods for measurement of *complex* Γ, which are described in the chapters which follow. In this case, three suitably chosen positions of the sliding short are sufficient to determine the circular locus. An alternative, and in many applications preferred, option is to use three 'off-set' shorts or reactances.

As an alternative[37] to the measurement of $\Gamma_{g(max)}$ (or $\Gamma_{l(max)}$) (which presupposes the availability of a technique for the

---

[37] In contrast to the *measurement* thereof, the realisation of a termination for which $\Gamma_g = \Gamma_{g(max)}$ does belong to the family of terminal invariant methods.

measurement of Γ) there are applications which require a termination for which $\Gamma_g = \Gamma_{g(max)}$ (or $\Gamma_g' = 0$). This may be recognised and realised as follows.

Referring to Fig. 15.10, the measurement system of Fig. 15.8 has been modified by the addition of the tuner T_X, although the same set

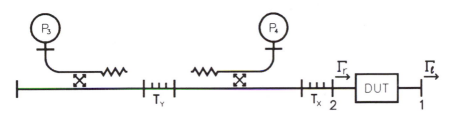

Fig. 15.10 *Measurement system for obtaining* $\Gamma_g' = 0$

of parameters, $A \cdots R_{c2}$, are used to describe the measurement system. Let the tuner T_Y be first adjusted for $R_{c2} = 0$ and then T_X for $R_{c1} = 0$.[38] As a consequence of the first of these adjustments, one has, from eqn. 14.17

$$BD^* - AC^* = 0 \qquad (15.70)$$

The parameters of the augmented network, which is obtained in Fig. 15.10 by a shift of the reference terminals from '2' to '1' (and thus include those of the DUT), may be obtained by comparing eqns. 15.45 and 15.48. The counterpart of eqn. 15.70, which follows from the adjustment of T_X, is

$$(Ab + B)(Cb + D)^* - (Aa + Bc)(Ca + Dc)^* = 0 \qquad (15.71)$$

Equation 15.70 may now be solved for B and substituted in eqn. 15.71. This leads to

$$(b^* - a^*c)\left(\frac{C^*}{D^*}\right)^2 + \left[1 + |b|^2 - |a|^2 - |c|^2\right]\frac{C^*}{D^*} + (b - ac^*) = 0$$
$$(15.72)$$

---

[38]Ideally (i.e. if T_x is lossless) these adjustments will be independent of one another. In practise, it may be necessary to repeat the operation.

As explained in Chapter 6, the reflection coefficient, Γ_g, of the 'equivalent source', which is provided by the second coupler in Fig. 15.10, is given by

$$\Gamma_g = -\frac{C}{D} \tag{15.73}$$

After making this substitution in eqn. 15.72, dividing by $\left[1 - |c|^2\right]$ and making use of the identity

$$\left[|a|^2 - |b|^2\right]\left[1 - |c|^2\right] = |a - bc|^2 - |b - ac^*|^2 \tag{15.74}$$

one has

$$R_c\Gamma_g^2 - \left[1 - R^2 + |R_c|^2\right]\Gamma_g + R_c^* = 0 \tag{15.75}$$

where R and R_c are given by eqns. 9.4 and 9.5 and where r has been set equal to unity. Comparison of this result with eqns. 15.67 and 15.68 now indicates that the value of Γ_g, which has been realised in this way, is such that $\Gamma_g = d^*$. Thus, from eqn. 15.35, one has $\Gamma_g' = 0$. Finally, it is possible to realise a *passive* termination, of the same reflection, by replacing the coupler on the left in Fig. 15.10 by a termination, and then adjusting T_Y for a null at P_4, with excitation applied at terminal '2'. The theoretical basis for this procedure is explained in Chapter 6.

Applications

A further expansion of the circuit in Fig. 15.7 is provided by Fig. 15.11, where additional lossless elements (LL) have been introduced to account for the source and load reflection. As illustrated here, and in Fig. 15.4, a key feature of the system model is the introduction of lossless elements in such a way as to more easily differentiate between the *reflection* of energy, as contrasted with the *absorption* of it. Moreover, the model is such as to permit these lossless elements to be combined with those of an adjoining component to yield a single lossless element, and whose parameters then become the measurement objective. In eqns. 15.9 and 15.10, for example, it is only necessary to determine $|\Gamma_{g3}|$, rather than the *complex* Γ_g and Γ_l. This feature, in turn, provides the basis for the insensitivity to certain connector problems. To be more specific, to the extent that the region in the vicinity of the connector interface may be considered *lossless*, the terminal surface may remain undefined and the effects of impedance discontinuities are included

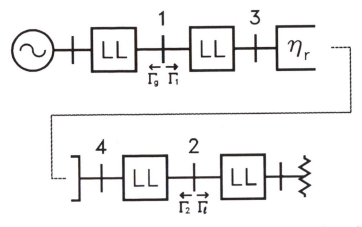

Fig. 15.11 Two-port model inserted between generator and termination

in the model. The connector parameter of primary interest, in this context, is its *repeatability*. The lossless assumption, of course, is never fully and in some cases may only be poorly realised, if at all. The implications of this, however, are that the assumed lossless region is reduced in size and this ultimately leads to a precise specification of the terminal surface. In the same context, the measurement methods which have been described rely heavily on either a sliding short or, alternatively, a variable or set of fixed reactances. In either case, the key element is the system response under three (as required to determine a circle) or more conditions under which there is zero power transfer. Although a lossless termination (reactance) is certainly the most convenient way of achieving this, it is not necessarily the only way. Thus these methods may continue to have an application in an environment where the reactance is not available. In more general terms, and returning to eqn. 14.19, one could conceivably determine the K_A, R and R_c if one were given the values of P_{gm}, P_4 and w_m, for four different terminations. The stipulation that three of the P_{gm} vanish is an operational convenience in that it simplifies the mathematical treatment and, ordinarily, the measurement technique as well. It is not an essential requirement of the method. The basic theory does not require the existence of these lossless elements as a condition for its validity, although the determination of the parameters on which it is based becomes more difficult in an environment where the approximations involved are no longer tenable. This area of potential application remains to be explored.

As of this writing, the major application of these methods has been in the context of power calibration and the associated simplification is well illustrated by eqns. 14.13 · · · 14.19. To be more specific, an evaluation of P_{gm} via eqn. 14.13 requires the measurement of *complex* Γ. This calls for a determination of the three *complex* parameters, A/D, B/D and C/D, which characterise the measuring instrument and which, in turn, involve the definitions of 'impedance' and 'matched termination'. By contrast, by use of the *power equation* methods (eqn. 14.19), this requirement has been replaced by that of determining a single complex parameter R_c and the real parameter, R. The definitions or conventions associated with *characteristic impedance* are *completely avoided*. As illustrated in this example, the measurement methods associated with the terminal invariant formulation tend to be easier to implement, although in an automated environment this frequently becomes a moot point. On the other hand, there is reason to anticipate that, as long as the ultimate measurement objective is a *terminal invariant* parameter, many of the measurement procedures associated with the automated network analyser, as described in the chapters which follow, may also satisfy the desired criteria, although unnecessary steps or constraints may be included in its calibration. This, however, needs to be confirmed on a case by case basis.

Although the *power equation* methods provide a simplified method for the evaluation of 'mismatch', (or deviations from the 'matched' condition) the techniques presuppose that the metrologist has access to all components of the system. If in Fig. 15.4, for example, the net effect of the cascade combination of lossless two-ports is to be experimentally determined, both source and termination must be available to the experimenter. A similar observation holds for Fig. 15.11 Alternatively, if one wishes to predict this mismatch interaction from the characteristics of the individual components, a more complete description, as found in the circuit or scattering equations, is required. This tends to restrict the application of these methods to the metrology laboratory.[39]

---

[39]For a description of some additional *power equation* methods see G. F. ENGEN: *'Theory of UHF and microwave measurements using the power equation concept'*, NBS Technical Note 637, April, 1973

Chapter 16

The automatic network analyser

The application of automation to the field of microwave measurements is perhaps best illustrated by the (vector) automated or automatic network analyser (ANA). The large reduction in measurement time and operator effort which it provides are well known, but this feature may tend to obscure a more fundamental aspect of its impact. Although techniques for measuring complex microwave impedance, scattering coefficients etc. have been known since the inception of the art, until the advent of the ANA much of microwave metrology was focused on the more easily determined scalar parameters such as power, attenuation and reflection coefficient magnitude. In the absence of phase information, it was common practise to determine 'worst case' limits for certain phase interactions which were labelled 'mismatch errors'.

As described in the preceding chapters, these techniques were largely based upon the use of detectors, P_3, P_4, which provided

$$P_3 = |Aa_2 + Bb_2|^2 \qquad (16.1)$$

and

$$P_4 = |Ca_2 + Db_2|^2 \qquad (16.2)$$

where a_2, b_2 are the wave amplitudes and $A \cdots D$ are system parameters. It will be recalled that the slotted line and reflectometer were based on certain idealized assumptions, namely $|A| = |B|$ or $B = C = 0$. The same general observation could also be made for numerous other techniques which have been described but failed to gain widespread acceptance. As a rule, in all these cases, the accuracy achieved is strongly contingent upon how well these idealized conditions are realised and for many years *the key to improved measurements was an improved item of microwave hardware*. The tuned reflectometer represents an exception. Here, as described in Chapter 12, the basic hardware imperfections may be eliminated by the use of appropriately placed tuning elements.[40] Although these methods proved both frequency sensitive and time consuming, and although no phase information was provided, this methodology served the art quite well for more than a decade. Moreover, the requirement for phase response was eliminated in certain problems by the use of the 'power equation' methods as described in an earlier chapter. However, with the pressure of broad-band measurement requirements, and with the advent of digital technology, the time was ripe for the emergence of the automatic network analyser.

The adjective 'automatic' has been applied to measurement schemes dating back five or more decades. Today, however, in the context of microwave metrology, the term 'automatic network analyser' (or ANA) usually represents a measurement concept associated with a digital computer which not only processes the measurement results, but also controls the data acquisition, source frequency and other parameters of the measurement system. In general these systems fall into two categories. In contrast to the 'vector' ANA, which is the primary subject of this chapter, the 'scalar' ANA typically represents a measurement technique which includes many of the same automation features, but with the important difference that the response is still based on eqns. 16.1 and 16.2. Here, although with improved designs the magnitude of the problem has been reduced, the errors introduced by the nonideal hardware still represent a major limitation to the attainable accuracy (and which may be determined by the methods described in Chapter 12).

By contrast, the vector ANA represents a complete shift in measurement strategy. In particular, with the addition of phase response in the detection system, the basic response, b_3, b_4, is now given by the equations

$$b_3 = Aa_2 + Bb_2 \qquad (16.3)$$

$$b_4 = Ca_2 + Db_2 \qquad (16.4)$$

where B and C no longer represent 'sources of error', but rather additional system parameters whose presence is explicitly recognised

---

[40]These are typically in the form of 'screws' or probes whose depth of penetration (and possibly longitudinal position) is adjustable.

and accounted for. In other words, the emphasis has shifted from an attempt to improve the hardware to a *smarter* use of that already in existence. As an additional bonus, the argument, as well as the magnitude of the microwave parameters being measured, e.g. complex reflection coefficient, is now readily obtained. In practise this phase response is typically achieved by the use of a heterodyne detection system, which may involve multiple frequency conversion, local oscillators, phase detectors etc. These elements, in turn, make a substantial contribution to the overall complexity of the system.

Returning to eqns. 16.3 and 16.4, their ratio may be written in terms of the more usual response, w

$$\frac{b_3}{b_4} = w = \frac{a\Gamma_l + b}{c\Gamma_l + 1} \qquad (16.5)$$

where $a = A/D$, $b = B/D$, $c = C/D$ and $\Gamma_l = a_2/b_2$. Solving this for Γ_l one has

$$\Gamma_l = \frac{w - b}{-cw + a} \qquad (16.6)$$

thus if a, b and c are known, Γ_l may be determined from the observed value of w.

At this point, it is of interest to make a further comparison between the microwave and lower frequency art. The role of Γ_l is analogous to impedance. At lower frequencies, the highest accuracy in impedance measurements is usually obtained by means of bridge techniques. Here, one or more of the bridge elements (or 'arms') may be adjusted to take on a series or continuum of known values. In use, these are adjusted to yield a null condition and the value of the unknown is determined in terms of the bridge parameters. At microwave frequencies, the measurement of Γ_l may also be considered to have been effected by bridge techniques, but where the bridge parameters are *fixed*. For this reason, and referring to eqn. 16.6, the measurement of an unknown Γ_l involves *both* the bridge parameters (a, b, c) *plus* the 'off-null' signal w. Where only the measurement of small values of $|\Gamma_l|$ are anticipated, it is obviously useful to choose these parameters such that $|b|$ is also small etc.

Calibration techniques

The application of the ANA calls first for its 'calibration' or an evaluation of the parameters a, b and c by which its operation is characterised. Since these have a frequency dependence, and their stability with time is frequently limited, there is a considerable premium placed on being able to do this both conveniently and accurately. One method of doing this calls merely for observing the response, w, for three different terminations, and for which Γ_l is known (impedance standards). From eqn. 16.6, this leads to three simultaneous equations which can be solved for a, b and c.

In a typical case, these may include a matched termination ($\Gamma_l = 0$), a 'flat' short and an 'offset' short. (Typically a quarter wavelength of transmission line terminated by a flat short) Whereas the first two may be nominally insensitive to frequency, this certainly is not true of the latter. This frequency sensitivity is avoided, and the requirement for multiple terminations of known reflection is eliminated by an alternative technique which will be next described. This calls for the use of a single termination of known reflection (usually a flat short) plus weakly and strongly reflecting sliding (or phaseable) terminations. These are usually in the form of a sliding 'load' and a sliding 'short', but neither of their reflection coefficients are actually required. (In particular, and in contrast to its application in the *power equation* context, the 'short' need not have a reflection of unity.)

Starting with eqns. 9.4 and 9.5 one has

$$R = \frac{|a - bc|\,|\Gamma_l|}{1 - |c|^2|\Gamma_l|^2} \tag{16.7}$$

$$R_c = \frac{b - ac^*|\Gamma_l|^2}{1 - |c|^2|\Gamma_l|^2} \tag{16.8}$$

where R and R_c are, respectively, the radius and centre of the circular locus of the response w to the sliding termination, the magnitude of whose reflection is $|\Gamma_l|$. Since $|\Gamma_l|$ is (usually) unknown, it is desirable to eliminate it. This may be accomplished by first solving eqn. 16.8 for $|\Gamma_l|^2$ and substituting in the denominator (only) of eqn. 16.7. The resulting expression is then solved for $|\Gamma_l|$, squared and substituted back into eqn. 16.8. This leads to

$$(b^* - R_c^*)(1 - R_c c/a) = R^2 c/a \qquad (16.9)$$

Let R and R_c be the parameters associated with the strongly reflecting termination, (sliding 'short') and let r and r_c represent those for the 'weak' reflection. By analogy with eqn. 16.9 one has

$$(b^* - r_c^*)(1 - r_c c/a) = r^2 c/a \qquad (16.10)$$

Equations 16.9 and 16.10 may now be solved for b and c/a. First, the elimination of c/a, between eqns. 16.9 and 16.10, yields a quadratic equation in $(b - r_c)$

$$(b - r_c)^2 (R_c^* - r_c^*) + (b - r_c)\left[R^2 - r^2 - |R_c - r_c|^2\right] + r^2(R_c - r_c) = 0. \qquad (16.11)$$

and the solution for b is given by[41]

$$b = r_c + \frac{2r^2(R_c - r_c)}{H + \left[H^2 - 4|R_c - r_c|^2 r^2\right]^{\frac{1}{2}}} \qquad (16.12)$$

where

$$H = R^2 - r^2 - |R_c - r_c|^2 \qquad (16.13)$$

Provided that $|c/a|$, and that the ratio of the reflection provided by the sliding 'load' to that of the sliding 'short', are both small, the magnitude of the second term in eqn. 16.12 is approximately $r^2|c/a|$ and is frequently small enough to be neglected. Thus, to a good approximation, $b \approx r_c$. After b has been determined, eqn. 16.9 may be solved for c/a,

$$c/a = (b^* - R_c^*)/(R^2 - |R_c|^2 + b^* R_c) \qquad (16.14)$$

Returning to eqn. 16.6, this may be written

$$\Gamma_l = \frac{1}{a}\left(\frac{w - b}{1 - wc/a}\right) \qquad (16.15)$$

---

[41] This root is the smaller in magnitude of the two. Except for gross departures from the usual design objectives, this will always be the root of interest.

thus the determination of b and c/a, which is provided by the sliding terminations, permits the determination of Γ_l apart from an unknown factor, $(1/a)$. Solving eqn. 16.15 for a, one has

$$a = \frac{1}{\Gamma_l}\left(\frac{w-b}{1-wc/a}\right) \tag{16.16}$$

To complete the calibration, it is only necessary to observe the response for a termination of known Γ_l. In practise a 'flat' short is frequently used, (for which ideally $\Gamma_l = -1$) but there is also a substantial precedent for the use of a so-called 'quarter wave' short, which is a quarter wavelength of line terminated by a short for which, ideally, $\Gamma_l = 1$. The advantage of the quarter wave over a flat short is that the longitudinal component of current is eliminated at the connector interface and with it the sensitivity to the series resistance which is associated with nonideal connector performance. On the other hand, the quarter wave short has a frequency sensitivity which is not shared by the flat variety. In a coaxial line, an 'open' circuit may be achieved by the simple expedient of extending the outer conductor.[42]   This tends to yield a higher value for the magnitude[43] of the reflection, but there will be some uncertainty in its argument because of the fringing fields at the termination of the inner conductor.   Referring again to eqn. 16.16, it is possible to combine the better features of the short and 'open' in a two step procedure where the magnitude of a is determined from the response to the 'open', while the argument is determined by that to the short.

Finally, the application of the above results requires the determination of R, R_c, r and r_c. As a consequence of eqns. 16.7 and 16.8 one has for the n different positions of the sliding termination

$$|w_i - R_c|^2 = R^2, \quad i = 1 \cdots n \tag{16.17}$$

---

[42]Practical experience with this technique indicates that the 'flower' or coupling mechanism ordinarily associated with the inner conductor should be removed, although this is contrary to one's intuitive expectation in a region where the current must vanish.

[43]Theoretical studies have shown that the difference between the magnitude of this reflection and unity is typically a few parts in 10^4, which by comparison with the related connector imperfections may ordinarily be neglected. For a further discussion of this topic see P. I. SOMLO & J. D. HUNTER: 'Microwave impedance measurement', (Peter Peregrinus Ltd., 1985)

After letting x_i, y_i and R_{cx}, R_{cy} represent the real and imaginary w_i and R_c, respectively, this may be expanded to yield

$$2x_i R_{cx} + 2y_i R_{cy} - (R^2 - |R_c|^2) = x_i^2 + y_i^2 \qquad (16.18)$$

By inspection, eqn. 16.18 is *linear* in R_{cx}, R_{cy} and $(R^2 - |R_c|^2)$. Although three positions of the sliding termination are sufficient, it is desirable in practise to use additional ones which then provide an overdetermined solution. Although the exact positions for the sliding terminations need not be known, the resulting values of w should be well distributed around the circumference in order to assure a well conditioned solution. In principle this same technique may also be used to determine r and r_c, but if the reflection of the sliding load is small enough, $r \rightarrow 0$, and the circular locus may be masked by the presence of detector noise. This can lead to gross errors if the above procedure is attempted. In this case, a usually acceptable alternative is to take the average of the associated values of w as the estimate of b.

Applications

Returning to eqn. 16.15, the parameters a, b, c, have now been determined and the measurement of an unknown reflection, Γ_l, may be obtained from the ANA response, w. This is illustrated in

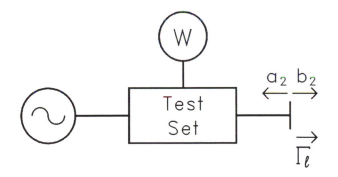

Fig. 16.1 Basic model for ANA

Fig. 16.1 where the measurement network (or 'test set'), which is characterised by a, b, c, and which provides the response, w, is typically a pair of directional couplers or a reflectometer. In order to

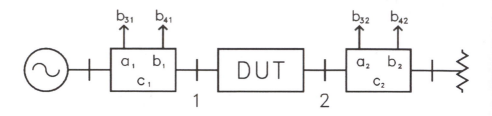

Fig. 16.2 Measurement of a two-port device

evaluate the parameters of a two-port the configuration in Fig. 16.2 may be used, which calls for a second test set.   The two-port scattering equations may be written[44]

$$b_1 = S_{11}a_1 + S_{12}a_2 \tag{16.19}$$

$$b_2 = S_{21}a_1 + S_{22}a_2 \tag{16.20}$$

As shown in Fig. 16.2, the test set at port 2 is terminated in a (matched) load.   Thus, ideally,[45] $a_2 = 0$ and $S_{11} = (b_1/a_1)$, while $S_{21} = (b_2/a_1)$. The first is provided by the first test set, while the second may be obtained by a simple extension of the techniques already described (including observations of the test set responses with the two-port removed). In a similar way, by interchanging the source and termination, one has $a_1 = 0$, so that S_{12} and S_{22} may be obtained. This now provides a complete determination of the two-port parameters. An alternative technique will be described in the following chapter.

---

[44]Note that this includes a redefinition of a_2, b_2 etc.

[45]The extension to the more general case where this idealisation is not satisfied is straightforward.

Chapter 17

An introduction to the

'Six-port' network analyser

In Chapter 16, the (vector) automatic network analyser (ANA) was introduced. Moreover, it was noted that a key element in its (usual) implementation is a detection system which provides *both* amplitude and phase response. This generally calls for a heterodyne detection system, possibly involving multiple frequency conversion and the associated local oscillators etc. In practise these components make a substantial contribution to the overall complexity and their realisation becomes increasingly difficulty as one moves into the millimeter wave region. Much of the earlier technology which it replaces was, by contrast, built on a much simpler detection system whose response was to amplitude (or power) only. Here the need for phase response can be avoided in certain problems by the use of techniques based on tuning transformers. However, these are both frequency sensitive and time consuming.

The so-called 'six-port' technique represents an attempt to combine the better features of the two technologies. In particular, the simplicity of the detection system is retained, while the phase response is achieved (and the tuning requirement eliminated) by the

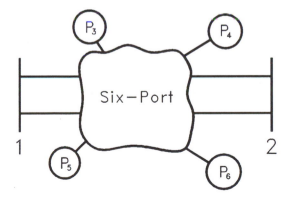

Fig. 17.1 The 'six-port' network analyser is based on a six-port junction

use of additional (usually a total of four) detectors. An additional advantage of the six-port technique is that in addition to network parameters, it includes the potential for making power measurements. The name 'six-port' was originally derived from the associated network where four of the 'ports' are terminated by these power detectors, while the remaining two provide for connection to the signal source and a 'test port' where the parameters of interest are measured. In retrospect, the choice of name is unfortunate since designs calling for three, five and more detectors are also of potential interest, although the major focus is on the use of four. By convention, and referring to Fig. 17.1, the port numbers 1 and 2 are usually reserved for the generator and test port, respectively, thus starting at port 3 one can write

$$P_3 = |Aa_2 + Bb_2|^2 \tag{17.1}$$

$$P_4 = |Ca_2 + Db_2|^2 \tag{17.2}$$

$$P_5 = |Ea_2 + Fb_2|^2 \tag{17.3}$$

$$P_6 = |Ga_2 + Hb_2|^2 \tag{17.4}$$

where $P_3 \cdots P_6$ are the sidearm responses, b_2 and a_2 are, respectively, the emergent and incident wave amplitudes at the test port and $A \cdots H$ are the six-port network parameters.

Linear model

Expanding (17.1) one has

$$P_3 = |A|^2|a_2|^2 + AB^*a_2b_2^* + A^*Ba_2^*b_2 + |B|^2|b_2|^2 \tag{17.5}$$

while similar expansions of eqns. 17.2 $\cdots$ 17.4 are also possible. By inspection, eqn. 17.5 is 'linear' in $|a_2|^2$, $a_2b_2^*$, $a_2^*b_2$ and $|b_2|^2$, thus the solution of eqns. 17.1 $\cdots$ 17.4 for these 'unknowns' is linear in $P_3 \cdots P_6$. Moreover, the *difference* between $|b_2|^2$ and $|a_2|^2$, and thus P_{net}, will also be linear in $P_3 \cdots P_6$. The reflection coefficient, Γ_l, may be written

$$\Gamma_l = \frac{a_2}{b_2} = \frac{a_2b_2^*}{|b_2|^2} \tag{17.6}$$

and combining these results one has

$$P_{net} = \sum_{i=3}^{6} Q_iP_i \tag{17.7}$$

and

$$\Gamma_l = \frac{\displaystyle\sum_{i=3}^{6} (u_i + jv_i)P_i}{\displaystyle\sum_{i=3}^{6} M_i P_i} \qquad (17.8)$$

where the P_i are the powers at the respective ports and where the Q_i, u_i, v_i and M_i are (real) system constants.

The network theory which leads to eqns. 17.7 and 17.8, however, does not account for certain redundancies which are inherent in the method. As will be shown below, the four power meter readings are connected by a quadratic equation such that three of them determine the fourth to the extent of a choice between two possible values. In practise it is possible to both improve and assess the measurement accuracy by taking explicit account of this feature.

In addition, the Q_i, u_i, v_i and M_i are not independent since collectively they represent a total of 16 constants, while 12 parameters suffice to describe the six-port measurement system. If not eliminated, this redundancy, in the parameters used to describe the system, may lead to unnecessary complications in the calibration procedure. Finally, while the application of eqns. 17.7 and 17.8 to certain (potentially useful) six-port configurations leads to a system of equations which are ill-conditioned, this problem disappears under the more complete treatment which follows.

Quadratic model

To continue, eqns. 17.1 $\cdots$ 17.4 may be factored to yield

$$P_3 = |A|^2 |b_2|^2 |\Gamma_l - q_3|^2 \qquad (17.9)$$

$$P_4 = |D|^2 |b_2|^2 |1 - \Gamma_l \Gamma_g|^2 \qquad (17.10)$$

$$P_5 = |E|^2 |b_2|^2 |\Gamma_l - q_5|^2 \qquad (17.11)$$

$$P_6 = |G|^2 |b_2|^2 |\Gamma_l - q_6|^2 \qquad (17.12)$$

where $q_3 = -B/A$, $\Gamma_g = -C/D$, $q_5 = -F/E$ and $q_6 = -H/G$.[46]

---

[46]It would be possible, of course, to write (17.10) in a manner similar to (17.9) etc. The chosen form avoids an indeterminate result when $C \to 0$.

At this point it is evident that the six-port can, at most, provide only the phase *difference* between a_2 and b_2 (as required for the argument of Γ_l). In addition, however, one also has the potential for measuring $|b_2|^2$ as required in power applications. (This added feature in not usually found in other types of automated network analysers.) Moreover, with each power detector is associated one real and one complex parameter ($|A|^2$ and q_3 in the case of P_3 etc.). Thus the six-port operation is completely characterised by four real and four complex parameters, (or 12 real parameters since a complex parameter is the equivalent of two real ones).

Equations $17.9 \cdots 17.12$ may now be considered an (over-determined) set in the real and complex measurands $|b_2|^2$ and Γ_l. The solution begins by eliminating $|b_2|^2$. This merely requires taking the ratio of three of the equations to the fourth which, for convenience, is eqn. 17.10. The ratio of eqn. 17.9 to eqn. 17.10 yields, after expansion, collecting terms and 'completing the square',

$$\left| \Gamma_l - \frac{q_3 - K^2 \Gamma_g^*}{1 - K^2 |\Gamma_g|^2} \right|^2 = \frac{K^2 |1 - q_3 \Gamma_g|^2}{\left[1 - K^2 |\Gamma_g|^2 \right]^2} \qquad (17.13)$$

where

$$K^2 = \frac{P_3}{P_4} \left| \frac{D}{A} \right|^2 \qquad (17.14)$$

In practise, the usual design objective calls for $\Gamma_g = 0$. If this has been realised, eqn. 17.13 becomes

$$|\Gamma_l - q_3|^2 = K^2 \qquad (17.15)$$

If the six-port parameters, which include A, D and q_3, have been determined (by a suitable calibration procedure) and P_3/P_4 has been observed, the only unknown in eqn. 17.15 is Γ_l. Although this equation fails to determine Γ_l, it is easily recognised, from Fig. 17.2, that the locus of possible values for Γ_l lie on a circle with centre at q_3. For reasons which will become apparent, it is convenient to initially assume that q_3 lies outside the unit circle as shown in Fig. 17.2 while, for a passive termination, Γ_l lies inside the circle.

Next, eqns. 17.10 and 17.11 may be combined and the radius of another circle, centred at q_5, and which also contains Γ_l, has been determined. The situation is now as shown in Fig. 17.3. Here Γ_l is determined by the intersection of the two circles. These circles,

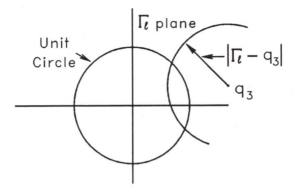

Fig. 17.2 Locus of possible values for Γ_l determined by P_3 and P_4

however, intersect in a pair of points. In this example the second point falls outside the unit circle and one is able to choose between the two solutions on the basis $|\Gamma_l| \leq 1$.

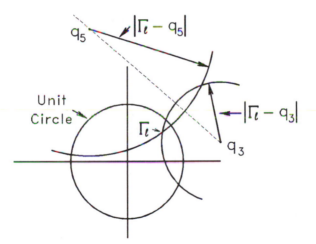

Fig. 17.3 Determination of Γ_l from the intersection of two circles

Thus far, no use has been made of P_6 and the system may be considered a five-port rather than six-port. Before introducing P_6, some additional observations on the five-port mode are of interest. As noted, this leads to a pair of values for Γ_l. Provided, however, that the straight line between q_3 and q_5 does not intersect the unit circle, one is assured that one of these roots will fall outside of it and (assuming a passive termination) may be rejected on this basis.

By further inspection of Fig. 17.3, one notes that, for the value of Γ_l used in this eqnample, the angle at which the circles intersect is rather small, thus the distance from the line which connects q_3 and q_5 to the circle intersection (Γ_l) will have a high sensitivity to errors in $|\Gamma_l - q_3|$ or $|\Gamma_l - q_5|$. In the parallel direction, the sensitivity is appreciably less. Over the range of possible choices for Γ_l and if Γ_l moves around the perimeter of the unit circle, for example, one can expect a considerable variation in these sensitivities or expected errors in a practical measurement system. Although the five-port measurement concept is technically viable, the primary purpose of this discussion has been to prepare one to better appreciate the benefits of the six-port over the five-port approach. Because these improvements are substantial, there has been little interest in the five-port.

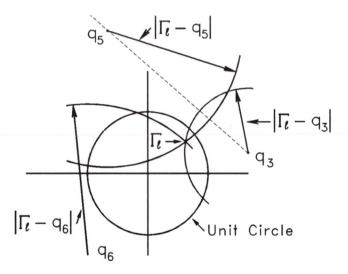

Fig. 17.4 An improved determination of Γ_l from the intersection of three circles

To continue, q_6 is chosen as shown in Fig. 17.4 and $|\Gamma_l - q_6|$ is determined from eqns. 17.10 and 17.12. This provides a third circle on which Γ_l must lie and which (ideally) passes through an intersection of the other two circles as shown in Fig. 17.4. In practise, because of measurement errors, the three circles will not intersect in a point and it is possible to infer the power meter error from this intersection failure.[47] Moreover, it is intuitively obvious that this additional detector has substantially enhanced the accuracy to which Γ_l is determined and the double root ambiguity has been resolved. It is no longer required that the line connecting q_3 and q_5 lies outside the unit circle.

In the more general case, where $\Gamma_g \neq 0$, inspection of eqn. 17.13 shows that the locus of Γ_l is still a circle, but where the center, R_c, as well as the radius, R, is a function of the power ratio. One now has

$$R = \frac{K\,|\,1 - q_3\Gamma_g\,|}{1 - K^2\,|\,\Gamma_g\,|^2} \tag{17.16}$$

$$R_c = \frac{q_3 - K^2\,\Gamma_g^*}{1 - K^2\,|\,\Gamma_g\,|^2} \tag{17.17}$$

A comparison of these results with eqns. 9.4 and 9.5 indicates that the relationship between R and R_c is governed by the linear fractional transform where the roles of a, b, c and r are provided by 1, q_3, Γ_g and K, respectively. Thus, as is evident from eqn. 17.13, the solution to the general problem can still be represented graphically as the intersection of three circles, but where R_c is no longer a constant but a function of K^2 and thus the item being measured. As a result, an additional element of substantial complexity has been introduced.

It is both possible and convenient, however, to retain the physical insight which is provided by the simple graphical model, (where the circle centres are constant) by an alternative two step solution. This begins with a return to four-port (or ANA) theory in which

$$b_3 = Aa_2 + Bb_2 \tag{17.18}$$

$$b_4 = Ca_2 + Db_2 \tag{17.19}$$

[47]G. F. ENGEN: 'On-line accuracy assessment for the dual six-port ANA: background and theory', *IEEE Trans. Instrum. Meas.*, IM-36, pp. 501-506, June 1987

R. M. JUDISH and G. F. ENGEN: 'On-line accuracy assessment for the dual six-port ANA: statistical methods for random errors', *IEEE Trans. Instrum. Meas.*, IM-36, pp. 507-513, June 1987

and letting

$$\frac{b_3}{b_4} = w = \frac{a\Gamma_l + b}{c\Gamma_l + 1} \tag{17.20}$$

where $a = A/D$, $b = B/D$ and $c = C/D$.

The solution now first calls for an evaluation of w,

$$w = F\left(\frac{P_3}{P_4}, \frac{P_5}{P_4}, \frac{P_6}{P_4}\right) \tag{17.21}$$

and following this, from eqn. 17.20,

$$\Gamma_l = \frac{w - b}{-cw + a} \tag{17.22}$$

With reference to eqn. 17.21, whose explicit form will be developed below, it is convenient to think of P_3/P_4, P_5/P_4 and P_6/P_4 as representing a point in a three-dimensional 'P-space'. Thus eqn. 17.21 represents a 'P-space to w-plane transformation'. Equations 17.20 and 17.22 will, of course, be recognised as the linear fractional or 'w-plane to Γ_l-plane transformation', whose role in the existing art has been covered in the earlier chapters. It is convenient to interpret eqn. 17.21 as having effected a 'six-to-four-port reduction' and also providing a complex ratio detector to which existing reflectometer theory is applicable. In general, the six-port may thus be applied in any measurement context where a four-port reflectometer and complex ratio detector would prove useful. In addition, and as noted above, because the individual sidearm signal levels are also obtained, the device is immediately applicable to power measurement applications as well.

In order to put eqn. 17.21 in explicit form, it is convenient to begin by solving eqns. 17.18 and 17.19 for a_2 and b_2 which are then substituted in eqns. 17.3 and 17.4. Recalling also the definition of w, this leads to

$$|w|^2 = P_3/P_4 \tag{17.23}$$

$$|w - w_1|^2 = gP_5/P_4 \tag{17.24}$$

$$|w - w_2|^2 = hP_6/P_4 \tag{17.25}$$

where

$$w_1 = \frac{A}{D} \frac{(q_5 - q_3)}{(1 - q_5 \Gamma_g)} \tag{17.26}$$

$$w_2 = \frac{A}{D} \frac{(q_6 - q_3)}{(1 - q_6 \Gamma_g)} \tag{17.27}$$

$$g = \left| \frac{A}{E} \right|^2 \cdot \left| \frac{1 - q_3 \Gamma_g}{1 - q_5 \Gamma_g} \right|^2 \tag{17.28}$$

$$h = \left| \frac{A}{G} \right|^2 \cdot \left| \frac{1 - q_3 \Gamma_g}{1 - q_6 \Gamma_g} \right|^2 \tag{17.29}$$

Equations 17.23 $\cdots$ 17.25 are in the same form as eqn. 17.15, but with the important result that the circle centres are now constant. As a further detail, it is evident from eqns. 17.24 and 17.25 that, at most, only the difference between the arguments of w and w_1 or w_2 can be obtained. For this reason it is possible, and has proven convenient, to include an arbitrary phase factor in w_1 such that, by definition, w_1 is real and positive.[48] At this point, eqns. 17.23 $\cdots$ 17.25 may be solved for w by first expanding eqns. 17.24 and 17.25 and then subtracting eqn. 17.23. This eliminates the quadratic terms and leaves a pair which are linear in the real and imaginary components of w whose solution is given by

$$w_x = \frac{P_3 - gP_5 + w_1{}^2 P_4}{2 w_1 P_4} \tag{17.30}$$

$$w_y = \frac{P_3 - hP_6 + (|w_2|^2 - 2w_x w_{2x})P_4}{2 w_{2y} P_4} \tag{17.31}$$

Finally, this result may be substituted in eqn. 17.22 to obtain Γ_l.

Ideally, as shown in Fig. 17.4 the three circles intersect in a point. In practise, because of measurement error, the situation is as

---

[48]This provides an implicit determination of the argument of A/D via eqn. 17.26.

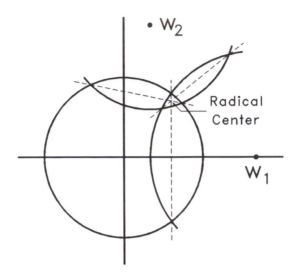

Fig. 17.5 The 'linear' solution of the six-port equations yields the intersection of the common chords.

shown in Fig. 17.5. Returning to eqn. 17.30, this is the equation of a straight line which passes through the points of intersection of eqns. 17.23 and 17.24. Equation 17.30 thus represents the common chord of the circles centred at the origin and w_1. In a similar manner, the common chord of the circles centred at the origin and w_2 is given by eqn. 17.31. A theorem from geometry states that the common chords of three circles intersect in a point which is, by definition, their 'radical center'. In general eqns. 17.30 and 17.31 or eqn. 17.8, thus yield the co-ordinates of the radical center. While the foregoing method of dealing with the problem of nonintersecting circles is the simplest from a computational point of view, it fails to utilise the inherent redundancy and its associated potential for an accuracy assessment.

In addition to representing the solution via the intersection of three circles, a further geometric interpretation is also of interest. With respect to eqns. 17.23 ··· 17.25, this set is overdetermined in that one has three equations from which to determine w_x and w_y. Alternatively, from Fig. 17.4 it is evident that the radii of any two of the circles determine that of the third to the extent of a choice between two possible values. It is possible to eliminate w from eqns. 17.23 ··· 17.25 by taking the solutions for w_x and w_y, given by eqns. 17.30 and 17.31, then substituting back in eqn. 17.23. This leads to

$$dx^2 + ey^2 + fz^2 + (f-d-e)xy + (e-d-f)xz + (d-e-f)yz$$

$$+ d(d-e-f)x + e(e-d-f)y + f(f-d-e)z + def = 0 \quad (17.32)$$

where

$$x = P_3/P_4 \quad (17.33)$$

$$y = gP_5/P_4 \quad (17.34)$$

$$z = hP_6/P_4 \quad (17.35)$$

$$d = |w_1-w_2|^2 \quad (17.36)$$

$$e = |w_2|^2 \quad (17.37)$$

$$f = |w_1|^2 \quad (17.38)$$

As expected, eqn. 17.32 is quadratic in x, y, z, (or the power ratios). Equation 17.32 is that of an elliptic paraboloid which, as

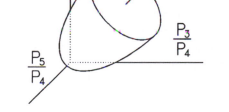

Fig. 17.6 The power ratios lie on an elliptic paraboloid in 'P-space'.

shown in Fig. 17.6, is confined to the first octant ($x \geq 0$, $y \geq 0$, $z \geq 0$) in xyz or 'P-space' and which is tangent to the planes $x = 0$, $y = 0$ and $z = 0$. The set of possible values for the power ratios are thus constrained to lie on this surface in P-space. In addition to providing added insight as to the nature of the constraint, this result also plays a key role in certain methods for calibrating the six-port whose description will be included in the material to follow.

The dual six-port

With minor extensions, the foregoing will make possible the measurement of power as well as reflection coefficient. Another major application is in the measurement of two-port parameters, e.g. attenuation etc. Here the 'dual six-port' has proven useful. As shown

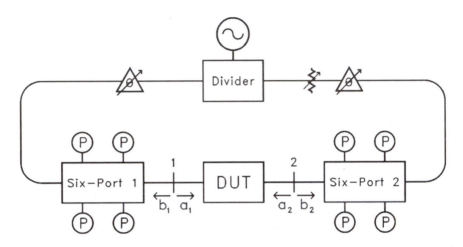

*Fig. 17.7 The **dual** six-port permits the measurement of two-port devices.*

in Fig. 17.7, this involves a pair of six-ports, each of whose inputs are connected, via a dividing network, to a common signal source. The two-port, which is to be measured, is inserted between the two six-ports. The dividing network includes provision for adjusting the phase and amplitudes of the signals incident upon the two-port, while the resulting response, at each end, is measured by the six-ports. Given the response from several such measurements, it is possible to obtain the two-port parameters as follows.

Referring again to Fig. 17.7, the incident and emergent wave amplitudes will be denoted by a_1, b_1 and a_2, b_2 at ports 1 and 2 of the two-port, respectively. (Note that this represents a change in the definition of a_2 and b_2 as previously used.) The two-port scattering equations may be written

$$b_1 = S_{11}a_1 + S_{12}a_2 \tag{17.39}$$

$$b_2 = S_{21}a_1 + S_{22}a_2 \tag{17.40}$$

Dividing eqns. 17.39 and 17.40 by a_1 and a_2, respectively, then eliminating the ratio a_2/a_1 between them, yields

$$(S_{12}S_{21} - S_{11}S_{22}) + \rho_2 S_{11} + \rho_1 S_{22} = \rho_1 \rho_2 \qquad (17.41)$$

where

$$\rho_i = b_i/a_i, \qquad i = 1, 2 \qquad (17.42)$$

In eqn. 17.41, ρ_1 and ρ_2 are the reflection coefficients as observed by six-ports 1 and 2. By changing the parameters of the dividing network, different sets of values for ρ_1 and ρ_2 may be obtained. These are then measured by the respective six-ports and substituted in eqn. 17.41. This yields a system of equations which are linear in S_{11}, S_{22} and $(S_{12}S_{21} - S_{11}S_{22})$. After solving these, $S_{12}S_{21}$ may be easily obtained as well. Provided that the two-port is reciprocal, $(S_{12} = S_{21})$, the foregoing is sufficient to determine the complete set of scattering parameters except for a 180° ambiguity in the argument of S_{12}. In many applications this is unimportant, but if necessary it can be resolved by further extensions of the method.

Dynamic range considerations

In an application of this type, the six-port provides a dynamic range for attenuation measurement which, if state-of-the-art bolometric power meters are used as the detectors, may be as large as 60 dB. On the other hand, the dynamic range of the bolometric technique, itself, is typically limited to 20 dB or less. An intuitive understanding of the basis for this may be obtained by starting with eqn. 17.41 which may be rewritten

$$S_{12}S_{21} = (\rho_1 - S_{11})(\rho_2 - S_{22}) \qquad (17.43)$$

Next, referring again to Fig. 17.7, it is convenient to consider the signal b_1 as the *vector* sum of a 'reflected' (S_{11}) component, due to a_1, and a 'transmitted' (S_{12}) component due to a_2. The phase shifter, which is part of the dividing network, makes possible the adjustment of the phase of a_2 with respect to a_1. This, in turn, leads to changes in the observed b_1. The ability to measure small values of $|S_{12}S_{21}|$ is thus closely related to the ability to recognise small changes in the observed ratio b_1/a_1 and ultimately in the power meter readings, $P_3 \cdots P_6$. Because the bolometric technique is already a differential measurement scheme, it is ideally suited for this purpose.[49] For 'six-port 1' in Fig. 17.7, the detector response, P_i, may be written in the alternative form

$$P_i = |A_i a_1 + B_i b_1|^2 = |C_i a_1 + D_i a_2|^2 \qquad (17.44)$$

where (in Fig. 17.7) $C_i = A_i + B_i S_{11}$ and $D_i = B_i S_{12}$. The term $C_i a_1$ thus includes contributions from both a_1 and that component of b_1 which is due to reflection (S_{11}) while $D_i a_2$ represents the transmitted component due to a_2, (or S_{12}). For large values of attenuation, $|C_i a_1| \gg |D_i a_2|$. Expanding eqn. 17.44 one has

$$P_i = |C_i a_1|^2 + 2|C_i D_i a_1 a_2|\cos\theta + |D_i a_2|^2 \qquad (17.45)$$

where θ is determined primarily by the setting of the phase shifter in the dividing network. To a first approximation, P_i is the sum of $|C_i a_2|^2$, which is nominally constant, plus a term which is *linear* in $|a_2|$ or $|S_{12}|$. The operation is thus similar to homodyne detection. For these reasons, a large dynamic range in the detectors is not required.

Calibration techniques

In common with other network analysers, an application of the six-port requires a 'calibration' or a determination of the parameters by which it is characterised. Although there is some reason to anticipate that the six-port will prove an exception, stability of the calibration (with time) is not ordinarily a characteristic feature of the ANA operation. Thus there is a substantial premium placed on being able to effect the required calibration quickly, conveniently and accurately. Moreover, in the primary laboratory environment, it is ordinarily desirable to keep the number of 'standards', terminations or other artifacts of 'known' characteristics to a minimum.

Although the number of parameters to be determined is nominally twice that for other network analysers, a number of methods which satisfy these criteria do exist. In particular it is possible to exploit the geometric interpretation which was introduced

---

[49] Although a number of six-port systems have been implemented which use diode detectors, their performance is generally inferior to those based on the use of bolometric detection. On the other hand, the reduced power requirement, associated therewith, is certainly an attractive feature where the ultimate in accuracy is not required.

in an earlier paragraph and make a determination of the five real parameters $(d \cdots h)$ which characterise the six-to-four-port reduction by use of eqn. 17.32 in conjunction with the observed response to a set of five or more[50] different excitation conditions and in which $d \cdots h$ are the unknowns. Here, the explicit characteristics of the terminations or other methods of achieving the different excitations can remain unknown except as required to assure a good distribution of points over the paraboloid surface or as required for a well conditioned solution. Once this set of parameters has been obtained, the six-port has been reduced to a 'four-port' with a complex ratio detector and any of the existing four-port calibration methods, which lead to the complex a, b, c, may be applied.[51] In general it is possible, and usually desirable, to include the set of power meter responses which are associated with this part of the calibration in the data set which is used to determine $d \cdots h$. The possible methods for calibrating the reduced 'four-port' include, of course, the sliding termination techniques described in an earlier chapter. It is of further interest to note that the associated data (for a sliding termination) lie in a plane[52] whose intersection with the paraboloid surface thus yields an ellipse in P-space. Calibration techniques which exploit this feature have also been described.[53]

---

[50] Equation (17.32) is *cubic* in the 'unknowns' $d \cdots h$ and an iteration procedure is required. In order to obtain an 'initial estimate' it is convenient to begin with the (linear) equation of a general quadric in three-dimensional space which involves nine parameters and sets of excitation conditions.

[51] The complex $a \cdots c$ and real $d \cdots h$ represent 11 parameters. The remaining or 12th parameter is only required in power measurement applications.

[52] This may be demonstrated as follows: In addition to satisfying eqns. 17.23 $\cdots$ 17.25, the observed w must lie on a circle in the w-plane, and thus satisfy an equation of the form $| w - R_c |^2 = R^2$. After eliminating $| w |^2$ between this and eqn. 17.23 one is left with a *linear* equation in the real and imaginary components of w. In a similar manner, eqn. 17.23 may be used to eliminate $| w |^2$ from eqns. 17.23 and 17.24. This yields two additional, or a total of three, equations which are *linear* in w_x and w_y and which are consistent if and only if there is a *linear* relationship among the P_i. Thus the P_i lie in a plane.

[53] G. F. ENGEN: 'Calibrating the six-port reflectometer by means of sliding terminations', *IEEE Trans. Microwave Theory Tech.*, MTT-26, pp. 951-957, Dec. 1978

The 'TRL' calibration technique

Although the foregoing methods are also applicable in the dual six-port environment, the availability of the second six-port makes possible the through-reflect-line (TRL) calibration technique which is particularly attractive. As shown in Fig. 17.8, and in keeping with

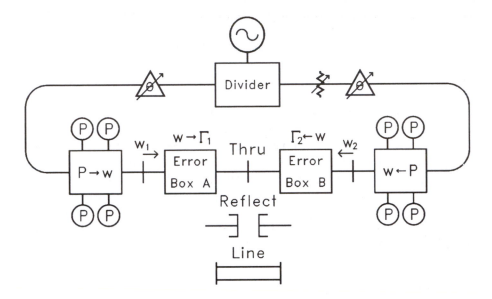

Fig. 17.8 Block diagram of 'Through-Reflect-Line' Calibration Technique

eqns. 17.20 ⋯ 17.22, the parameters which characterise the six-to four-port reduction, (or $P \rightarrow w$ transform) are first determined as already described. This permits the measurement of w, whose relationship to Γ may be 'modeled' by a fictitious two-port 'error box'. (See eqn. 17.20.) The method then[54] calls for observing the system response with the two six-ports connected together (*Through*), next with the same 'unknown' termination (usually a flat 'short') connected in turn to each of them (*Reflect*) and finally with an arbitrary and unknown (but different from 180°!) length of line inserted between them (*Line*). With the help of eqn. 17.41, it is then possible to determine the parameters of the (*Through*) 'two-port' which is comprised of error boxes 'A' and 'B' in cascade and similarly for the (*Line*) 'two-port' where an arbitrary length of line has been inserted

---

[54]In reality, this data has already been collected and utilized in the determination of the parameters which characterize the $P \rightarrow w$ transform.

between them. In keeping with eqn. 4.9, let the cascading matrices of the error boxes **A** and **B** be denoted by $\mathbf{R}_a$ and $\mathbf{R}_b$, while $\mathbf{R}_t$ represents their cascade 'through' connection. Then

$$\mathbf{R}_t = \mathbf{R}_a\mathbf{R}_b \qquad (17.46)$$

while if $\mathbf{R}_d$ represents the 'line' combination,

$$\mathbf{R}_d = \mathbf{R}_a\mathbf{R}_l\mathbf{R}_b \qquad (17.47)$$

where $\mathbf{R}_l$ represents the line which has been inserted.

Eliminating $\mathbf{R}_b$ between eqns. 17.46 and 17.47 yields

$$\mathbf{TR}_a = \mathbf{R}_a\mathbf{R}_l \qquad (17.48)$$

where

$$\mathbf{T} = \mathbf{R}_d\mathbf{R}_t^{-1} \qquad (17.49)$$

and which may be obtained from the parameters of the through and line 'two-ports'.

If γ and l represent the propagation constant and length of the line, and assuming it to be nonreflecting, one has

$$\mathbf{R}_l = \begin{pmatrix} e^{-\gamma l} & 0 \\ 0 & e^{\gamma l} \end{pmatrix} \qquad (17.50)$$

Finally, the elements of $\mathbf{R}_a$ and $\mathbf{T}$ will be represented by r_{ij} and t_{ij}, respectively. Expansion of eqn. 17.48 gives

$$t_{11}r_{11} + t_{12}r_{21} = r_{11}e^{-\gamma l} \qquad (17.51)$$

$$t_{21}r_{11} + t_{22}r_{21} = r_{21}e^{-\gamma l} \qquad (17.52)$$

$$t_{11}r_{12} + t_{12}r_{22} = r_{12}e^{\gamma l} \qquad (17.53)$$

$$t_{21}r_{12} + t_{22}r_{22} = r_{22}e^{\gamma l} \qquad (17.54)$$

The ratio of eqn. 17.51 to eqn. 17.52, and of eqn. 17.53 to eqn. 17.54 yields

$$t_{21}(r_{12}/r_{22})^2 + (t_{22} - t_{11})(r_{12}/r_{22}) - t_{12} = 0 \qquad (17.55)$$

and

$$t_{21}(r_{11}/r_{21})^2 + (t_{22} - t_{11})(r_{11}/r_{21}) - t_{12} = 0 \qquad (17.56)$$

By use of eqns. 4.12 $\cdots$ 4.14 this may be written

$$Cb^2 + (1 - A)b - B = 0 \qquad (17.57)$$

$$C(a/c)^2 + (1 - A)(a/c) - B = 0 \qquad (17.58)$$

where $a \cdots c$ are the parameters of error box **A** which relate w and Γ and $A \cdots C$ are the counterpart for the **T** matrix.

The parameters b and (a/c) are thus given by the solution to the same quadratic equation where the coefficients may be obtained from the **T** matrix. Ideally, $c = 0$, so that ordinarily $|b| \ll |a/c|$ and this may be used as the basis for root selection. The ratio of eqn. 17.54 to eqn. 17.52 gives

$$e^{2\gamma l} = \frac{Cb + 1}{B(c/a) + A} \qquad (17.59)$$

Returning to eqn. 17.46, this may be written

$$r_{22}\rho_{22} \begin{bmatrix} a & b \\ c & 1 \end{bmatrix} \begin{bmatrix} \alpha & -\gamma \\ -\beta & 1 \end{bmatrix} = t_{22} \begin{bmatrix} A & B \\ C & 1 \end{bmatrix} \qquad (17.60)$$

where $\alpha \cdots \gamma^{55}$ are the counterpart of $a \cdots c$ for error box **B** and where the interchange of β and γ and the sign reversal account for the 'reversed' direction of propagation associated with the cascade connection of error box **B**. After premultiplying eqn. 17.60 by $\mathbf{R}_a^{-1}$ and expanding it is easy to show that

$$\beta = \frac{C - Ac/a}{1 - Bc/a} \qquad (17.61)$$

$$\gamma/\alpha = \frac{b - B}{A - Cb} \qquad (17.62)$$

and

---

[55] This use of γ must not be confused with its prior use in the exponential associated with the line!!

$$a\alpha = \frac{A - Cb}{1 - Bc/a} \qquad (17.63)$$

As already noted, $A \cdots C$ are the experimentally determined parameters of the **T** matrix, while b and (c/a) are obtained from the solution of eqns. 17.57 and 17.58. Thus β, (γ/α) and $a\alpha$ may be obtained from eqns. 17.61 $\cdots$ 17.63.

To continue, from eqn. 17.22 one has

$$a = \frac{1}{\Gamma_l}\left(\frac{w_1 - b}{1 - w_1 c/a}\right) \qquad (17.64)$$

where w_1 is the observed response to the termination Γ_l. Thus if Γ_l is known, the calibration of error box **A** may be completed by use of eqn. 17.64 as also described in Chapter 16. (See the discussion following eqn. 16.16.) It is of interest to note that the parameters b and (c/a) are just those which were determined by the sliding terminations. However, the need for these has been eliminated, thus the scope of potential applicability for the TRL technique is far greater and includes lines of *solid* dielectric etc. In a similar manner, α may be determined from

$$\alpha = \frac{1}{\Gamma_l}\left(\frac{w_2 - \beta}{1 - w_2\gamma/\alpha}\right) \qquad (17.65)$$

which is the counterpart of eqn. 17.64 for error box **B**.

The need for a termination (Γ_l) of known reflection may be *eliminated (!!)*, however, by taking the ratio of eqn. 17.64 to eqn. 17.65. This gives

$$a/\alpha = \frac{(w_1 - b)(1 - w_2\gamma/\alpha)}{(w_2 - \beta)(1 - w_1 c/a)} \qquad (17.66)$$

and which may be combined with eqn. 17.63 to yield

$$a = \pm\left[\frac{(w_1 - b)(1 - w_2\gamma/\alpha)(A - Cb)}{(w_2 - \beta)(1 - w_1 c/a)(1 - Bc/a)}\right]^{\frac{1}{2}} \qquad (17.67)$$

and

$$\alpha = \frac{A - Cb}{a(1 - Bc/a)} \tag{17.68}$$

Apart from the choice of sign in eqn. 17.67, the requirement for a known value of Γ_l has been *eliminated*. As a practical matter, a nominal short continues to be a convenient choice for Γ_l, although in some applications an 'open' may provide better repeatability. In either case, a nominal value for the argument is usually available and, with the help of eqn. 17.64, this permits the proper choice of sign to be made.

In addition to its wider scope of applicability and the elimination of the requirement for a termination of known reflection, this technique has the important advantage of putting the role of the *uniform line* (to which microwave circuit theory owes its existence) into better focus. The inevitable dissipation associated with the line is explicitly recognised and accounted for. In addition to a determination of the complex $a \cdots c$, as required to characterise the six-ports, the method also provides the complex propagation constant of the line, its electrical length and the reflection coefficient of the 'unknown' termination which has been used in the technique.[56]

Application to power measurements

As noted in an earlier paragraph, the six-port technique includes the potential for making power measurement. In order to put this in more definitive form, it is convenient to begin with Fig. 17.9, which provides a more explicit indication of the boundary conditions associated with error box **A**. The wave amplitudes a_2 and b_2, which

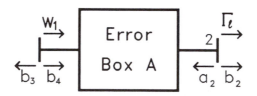

Fig. 17.9 Boundary conditions for error box A

[56]For further details see G. F. ENGEN and C. A. HOER: 'Through-reflect-line: An improved technique for calibrating the dual six-port automatic network analyzer', *IEEE Trans Microwave Theory Tech.*, MTT-27, pp. 1473-1477, Dec. 1979

describe the excitation state at the measurement plane, are related to b_3 and b_4 which obtain at the w plane by,

$$\begin{pmatrix} a_2 \\ b_2 \end{pmatrix} = \mathbf{R}_a^{-1} \begin{pmatrix} b_3 \\ b_4 \end{pmatrix} \qquad (17.69)$$

where

$$\mathbf{R}_a^{-1} = \frac{1}{r_{22}(a - bc)} \begin{pmatrix} 1 & -b \\ -c & a \end{pmatrix} \qquad (17.70)$$

The (net) power output from error box **A** is given by

$$P_2 = |b_2|^2 \left[1 - |\Gamma_l|^2 \right] \qquad (17.71)$$

where Γ_l may be obtained from eqn. 17.22, while from eqns. 17.69 and 17.70,

$$|b_2|^2 = \frac{|a_1(a - cw_1)|^2}{|r_{22}(a - bc)|^2} \qquad (17.72)$$

After making these substitutions and noting that $|b_4|^2 \equiv P_4$, eqn. 17.71 becomes

$$P_2 = \frac{P_4 |a - cw_1|^2}{|r_{22}(a - bc)|^2} \left[1 - \frac{|w_1 - b|^2}{|a - cw_1|^2} \right] \qquad (17.73)$$

The only system parameter which is undetermined at this point is $|r_{22}|$ which may be found by observing the system responses, P_4 and w_1, with P_2 indicated by a terminating type 'standard' power meter. After this has been done, the power delivered to other terminations may be found by use of eqn. 17.73.

Practical six-port circuits

Thus far, little has been said about the design of the six-port circuit. Referring to Fig. 17.4, the locations of the points q_3, q_5, q_6 are determined by the six-port parameters. There is also a fourth point ('q_4' or $1/\Gamma_g$) associated with the detector P_4, which is ideally at infinity. In addition there are certain scale factors associated with P_3, P_5, P_6 which combine with the q_i to make up the 11 parameters required to describe the six-port. The question of six-port design, however, revolves largely around the choice of the q_i. Although the six-port theory is, in principle, applicable to arbitrary designs, this does not mean that all designs are equally satisfactory!! In keeping with the discussion associated with eqns. 17.44 and 17.45, it is generally desirable to provide the detectors with a substantial component of the incident or emergent signal[57] ($A_i b_r$). As a general rule, in order to assure a well conditioned solution and achieve the best accuracy, design objectives call for placing the 'q' points in a symmetric position about the origin and giving them a nominal magnitude of 1.2-1.8.

A simple six-port circuit is shown in Fig. 17.10. Here a directional coupler is preceded by three nondirectional probes which,

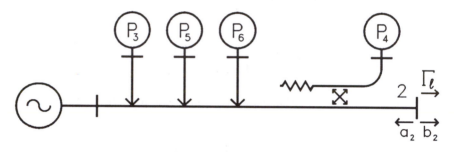

Fig. 17.10 A simple six-port circuit

for optimal performance, are spaced $\lambda/6$ apart. In this circuit, and the others which follow, P_4 is (ideally) proportional to $|b_2|^2$, thus q_4 is at infinity. As a result of the attenuation associated with the directional coupler, the probes will be coupled somewhat more strongly to the incident or emergent signal, b_2, than the reflected signal, a_2. The remainder of the q_i magnitudes are thus somewhat greater than unity whereas, for the assumed probe placement, the arguments differ by 120°. Since the wavelength, λ, is frequency dependent, the bandwidth of this design is rather limited, but it has proven useful in rectangular waveguides.

---

[57]This is in marked contrast to the *reflectometer* where it is ordinarily desirable to provide a detector which responds only to the *reflected* signal. In the six-port context, this calls for one of the 'q_i to equal zero. If this is done, however, the dynamic range requirements for the associated detector are greatly increased.

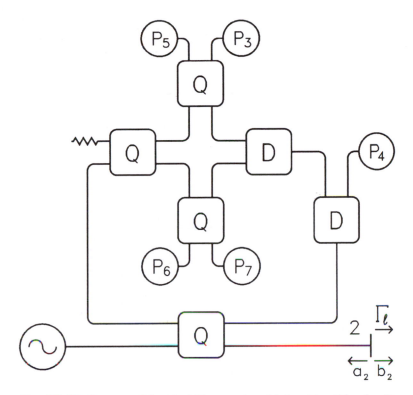

Fig. 17.11 Improved bandwidth may be obtained by this circuit.

In order to increase the bandwidth, the circuit of Fig. 17.11 may be employed.   Here 'D' and 'Q' represent, respectively, a power divider and a quadrature hybrid. This is, in reality, a seven-port. Apart from q_4, the q_i are on the perimeter of the unit circle and spaced 90° apart. A six-port may be obtained by replacing one of the power meters, P_3, P_5, P_6 or P_7, with a termination. Although this circuit has found substantial application as a 'six-port', the elimination of one of the q_i leaves a spacing of 90°, 90° and 180° between the remainder which is not very satisfactory. In this example, and the one which follows, it is possible to obtain bandwidths of 10-1 or 20-1 by the use of stripline components.

An alternative circuit is shown in Fig. 17.12. As indicated in Fig. 17.13, the angles between the q_i are 90°, 135° and 135°. Moreover, $|q_5| = |q_6| = 2$, and $|q_3| = \sqrt{2}$. If desired, these may be

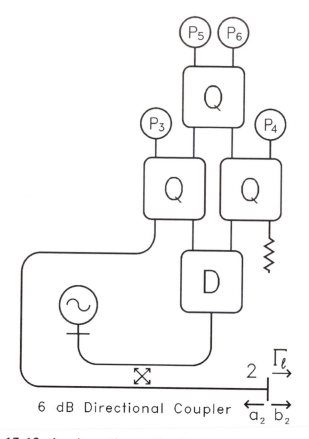

Fig. 17.12 An alternative to Fig. 17.11 using fewer hybrids

brought closer to the unit circle (by the factor $\sqrt{2}/2$) by changing the 6 dB coupler to a 3 dB one. As compared with that of Fig. 17.11, this circuit requires one less hybrid and provides a somewhat better disposition of the q_i.[58] Moreover, although a resistive termination is included, the circuit is inherently 'lossless' in that (ideally) none of the power reaches this port. Thus a better utilisation of the input power is realised. Assuming 20 mW of power at the input, 5 mW (or 1/4) of this reaches the measurement port. For a matched termination, this power will be absorbed while the remaining 3/4 will

---

[58]For a further discussion, including variants, see G. F. ENGEN: 'An improved circuit for implementing the six-port technique of microwave measurements', *IEEE Trans. Microwave Theory Tech.*, MTT-25, pp. 1080-1083, Dec. 1977

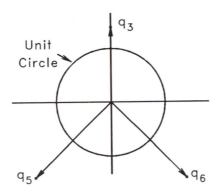

Fig. 17.13 Location of the 'q' points for the circuit of Fig. 17.12

be divided equally among the detectors $P_3 \cdots P_6$, resulting in a power level of 3.75 mW for each. If now P_4 is stabilised at this value, and a sliding short is connected to the measurement port, the value of P_3 will reach approximately 11 mW for certain short positions, while the maximums at P_5 and P_6 will be approximately 8.5 mW. The maximum dynamic range excursion at any detector is a nominal 15 dB which occurs at P_3.

   Another circuit is shown in Fig. 17.14. Here, the feature of interest is a five-port junction which has been combined with a directional coupler to provide the six-port circuit. Provided only that

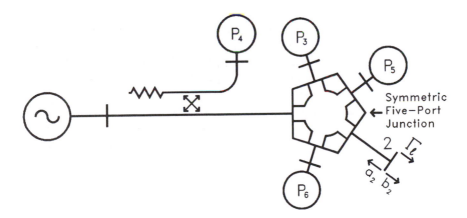

Fig. 17.14 A six-port circuit utilising a symmetric five-port junction

the five-port junction is *lossless, symmetric* and *matched*, it can be shown[59] that the resulting q_i are of magnitude *and spaced 120° apart!!*. In practise, although none of these criteria are completely satisfied, they may be closely approximated. The primary problem is with the third which calls for a reflection free input at one port when the remainder are connected to reflectionless terminations. Although this may be readily achieved at a single frequency, the bandwidth for this design is smaller than that provided by the circuits of Figs. 17.11 and 17.13. On the other hand, in stripline at least, it is much easier to implement.

The 'multi-state' reflectometer

As shown in Fig. 17.15, an interesting variant of the six-port concept is the 'multi-state' reflectometer developed at the Royal Signals and Radar Establishment (England). The circuit calls for the

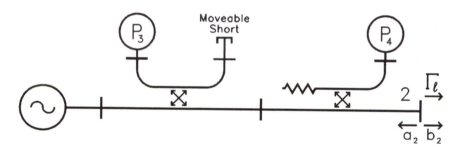

Fig. 17.15 The 'Multi-state' reflectometer

use of two directional couplers and two (rather than four) detectors. The coupler-detector combination on the right provides a (nominal) indication of the incident wave amplitude. The unique feature is associated with the coupler on the left. Here, the 'internal' termination on the fourth arm has been replaced by a moving short. In combination with the detector, P_3, this provides the equivalent of a nondirective moving probe. In operation, the ratios P_3/P_4 are observed for three positions of the moving short (which may be arbitrary, but must be repeatable). These three ratios are then the counterpart of the response of the conventional six-port. It is

[59]E. R. B. HANSON and G. P. RIBLET: 'An ideal six-port network consisting of a matched reciprocal lossless five-port and a perfect directional coupler', *IEEE Trans. Microwave Theory Tech.*, MTT-31, pp. 283-288, Mar. 1983

important to note that, as described in Chapter 12, the coupler arrangement is such that the relationship between P_4 and the test port signals is independent of the moving short. Compared with the more conventional six-port, fewer components are involved and less power is required. These are important practical considerations at millimeter wave frequencies.[60]

Comparison with other methods

As compared with other automated network analysers, the six-port is based on a much simpler detection system and this is of particular interest in the millimeter frequency region. (Systems operating at 400 GHz using quasi-optical techniques have been reported.) By contrast, the six-port detector response, $P_3 \cdots P_6$ requires a substantial amount of processing, although still well within the scope of a desk top computer. Since the detectors are broadband, there is less sensitivity to narrow band source jitter and noise, but an increased sensitivity to harmonics and wideband noise. Although the six-port model is a fairly complicated one, it also describes the associated hardware to a high degree of accuracy. Thus it is of particular interest in the standards laboratory environment. Here, the best accuracy is usually provided by thermistor type detectors. Unfortunately, however, these also require substantially more power than alternative detection schemes.

At the other end of the applications spectrum it appears that the six-port methods should also provide the basis for certain types of *in-situ* monitoring. This region of potential applicability is largely unexplored.

---

[60]L. C. OLDFIELD, J. P. IDE and E. J. GRIFFIN: 'A multi-state reflectometer', *IEEE Trans. Instrum. Meas.*, IM-34, pp. 198-201, June, 1985

Chapter 18

A brief survey of

Adapter evaluation techniques

Because the microwave art is characterised by the use of a variety of waveguides or transmission lines, the subject of adapters and their evaluation is an important one to the microwave metrologist. A typical problem is that of using a power or noise standard in a waveguide (for example) to calibrate an 'unknown' in coaxial line. Today, as will be described below, an explicit and complete adapter evaluation is possible using the automated vector network analysers. In addition, however, the prior art also includes techniques which provide an *approximate* elimination of the adapter parameters from the problem of interest. It is the purpose of this chapter to provide, with the help of specific examples, a brief survey of the existing methods.

Elimination of adapter parameters

As described in Chapter 13, the parameter of interest in a terminating power meter is ordinarily its (effective) efficiency.[61] Thus, given the efficiency, η_w, of a waveguide power 'standard', the problem is that of calibrating or obtaining the efficiency, η_c, of an 'unknown' coaxial power meter. Apart from the change in transmission line, this may be achieved by means of the techniques described in Chapter 14, for example. Moreover, if the adapter is *lossless* (which may be a fairly good approximation) the efficiencies, η_c or η_w, are not affected or altered by the addition of an adapter to their input. Thus the same techniques are potentially applicable. To a first approximation, the effect of adapter loss may be eliminated as follows.

The method calls for a pair of measurements (both involving the *same* adapter) which is first used to convert the coaxial meter to waveguide. The adapter-coaxial meter combination is then compared or calibrated against the waveguide standard. In the second step, the

---

[61]Since it is unimportant in what follows, the distinction between efficiency and *effective efficiency* will be dropped in this chapter.

waveguide standard becomes a coaxial one by use of the adapter and the calibration procedure repeated. In the absence of the adapter, let the result of the measurement or comparison procedure, m, be such

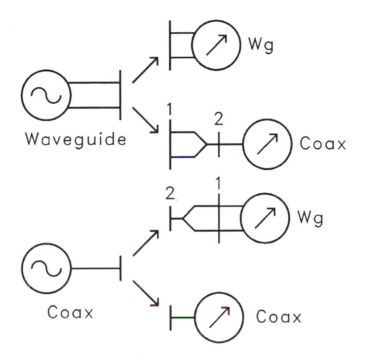

Fig. 18.1 *The effect of adapter losses may be approximately eliminated by the use of this technique.*

as to yield the ratio η_c/η_w ($m = \eta_c/\eta_w$). This, in turn, determines η_c, since by hypothesis η_w is known. Denoting the results of these two measurements, as shown in Fig. 18.1, by m_1 and m_2, one has

$$m_1 \;=\; \frac{\eta_{21}\eta_c}{\eta_w} \tag{18.1}$$

and

$$m_2 \;=\; \frac{\eta_c}{\eta_{12}\eta_w} \tag{18.2}$$

where η_{21} and η_{12} are, respectively, the adapter efficiencies in the two modes of operation. The geometric mean of m_1 and m_2 yields

$$\sqrt{m_1 m_2} = \frac{\eta_c}{\eta_w} \left(\frac{\eta_{21}}{\eta_{12}} \right)^{\frac{1}{2}} \tag{18.3}$$

which is the result of interest. In practise, $\eta_{12} \approx \eta_{21}$. Thus, eqn. 18.3 gives η_c/η_w with an error E which is given by

$$E = \left(\frac{\eta_{21}}{\eta_{12}} \right)^{\frac{1}{2}} - 1 \tag{18.4}$$

An expression for η_{21} is given by eqn. 5.13 with $\Gamma_l = \Gamma_c$ and that for η_{12} may be obtained by interchanging the subscripts and replacing Γ_c by Γ_w. If $S_{11} = S_{22}$ and $\Gamma_c = \Gamma_w$, then $\eta_{21} = \eta_{12}$. The result of interest thus tends to be favoured by the (usual) design objectives: $S_{11} = S_{22} = \Gamma_c = \Gamma_w = 0$. In addition, as the losses become small, the efficiencies tend to unity, irrespective of the deviations from the stated design objectives. This also tends to assure the equality. In a typical application, both sets of criteria are *approximately* satisfied, thus the error in assuming $\eta_{21} = \eta_{12}$ is usually of *second order*. A more complete evaluation of the error limits is obtained as follows.

Taking the ratio of eqns. 18.1 and 18.2, one has

$$\frac{m_1}{m_2} = \eta_{12} \eta_{21} \tag{18.5}$$

which provides a measure of the adapter losses. Let

$$\epsilon = \frac{1}{\eta_{12} \eta_{21}} - 1 \tag{18.6}$$

In addition, as shown in Fig. 18.2, let Γ_w and Γ_c represent the reflection coefficients of the waveguide and coaxial mounts and let Γ_1 be the reflection coefficient at terminal 1 of the adapter when terminal 2 is terminated by the coaxial mount, while Γ_2 is that at terminal 2 of the adapter when terminal 1 is connected to the waveguide mount. Finally, let $|\Gamma_a|$ represent the adapter reflection

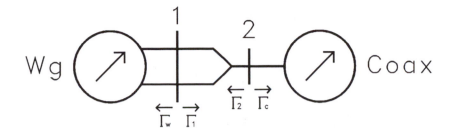

Fig. 18.2 Definitions of Γ_1, Γ_2, Γ_w and Γ_c

coefficient.[62] It has been shown[63] that approximate limits for E, for three different cases, are as follows:

Case I. The impedance conditions are completely arbitrary. Limits for E are given by

$$E_{max} = \frac{\epsilon}{2}\left[\,|\Gamma_w| + |\Gamma_c| + |\Gamma_a|\,\right] + \frac{\epsilon^2}{8} \qquad (18.7)$$

$$E_{min} = -\frac{\epsilon}{2}\left[\,|\Gamma_w| + |\Gamma_c| + |\Gamma_a|\,\right] - \frac{\epsilon^2}{8} \qquad (18.8)$$

Case II. It is assumed that Γ_2 and Γ_c are equal (in *both* amplitude and phase) but unknown. This presupposes the incorporation and use of a tuning transformer in one of the components (usually the adapter). The limits for E are now

---

[62] The adapter reflection coefficient magnitude, $|\Gamma_a|$, is that value which obtains at one side or port of the adapter with the other end connected to a matched (reflectionless) termination. It thus corresponds to the 'adapter VSWR'. In general, the value of $|\Gamma_a|$ measured at terminal 1 differs from that at terminal 2. For adapters of high efficiency, this difference is small, and vanishes as the adapter becomes lossless. Thus, for most practical purposes, the adapter may be regarded as being characterised by a single value of $|\Gamma_a|$.

The error expressions to be given for 'Case I' provide the correct limits if the value substituted for $|\Gamma_a|$ is the smaller of the two. Thus, if the larger value is used instead, somewhat wider limits will be obtained.

The failure to identify $|\Gamma_a|$ with either terminal is intentional in that this represents the most general case of practical significance. As will be shown, tighter limits of error result if $|\Gamma_a|$ is identified with one terminal or the other.

[63] G. F. ENGEN: 'Coaxial power meter calibration using a waveguide standard', *J. Res. NBS*, 70C, pp.127-138, Apr.-June, 1966

$$E_{max} = \frac{1}{2}\Big\{ |\Gamma_1| + |\Gamma_w| \Big\}^2, \quad \text{if } |\Gamma_1| + |\Gamma_w| \leq \frac{\epsilon}{2} \qquad (18.9)$$

$$E_{max} = \frac{\epsilon}{2}\Big\{ |\Gamma_1| + |\Gamma_w| \Big\} - \frac{\epsilon^2}{8}, \quad \text{if } |\Gamma_1| + |\Gamma_w| \geq \frac{\epsilon}{2} \qquad (18.10)$$

$$E_{min} = -\frac{\epsilon}{2}\Big\{ |\Gamma_1| + |\Gamma_w| \Big\} - \frac{\epsilon^2}{8} \qquad (18.11)$$

Case III. The reflection coefficients Γ_1 and Γ_w are equal and of known magnitude. The limits for E are now given by

$$E_{max} = 2|\Gamma_w|^2, \quad \text{if } |\Gamma_w| \leq \frac{\epsilon}{4} \qquad (18,12)$$

$$E_{max} = \epsilon|\Gamma_w| - \frac{\epsilon^2}{8}, \quad \text{if } |\Gamma_w| \geq \frac{\epsilon}{4} \qquad (18.13)$$

$$E_{min} = -\epsilon|\Gamma_w| - \frac{\epsilon^2}{8} \qquad (18.14)$$

The error limits for Case III may thus be obtained from those of Case II by letting $|\Gamma_1| = |\Gamma_w|$.

The foregoing techniques provide a method of minimizing the adapter error in an environment where a complete characterisation of the adapter or a determination, of its parameters, may not be convenient or possible.[64] With the advent of the ANA, however, this is often feasible and, in turn, makes possible an explicit determination of E. This will be the subject of the section which follows.

Evaluation of adapter parameters

Referring to Fig. 18.3, a reflectometer (or ANA) will be assumed whose test or measurement port is connected to the adapter of

---

[64]For additional applications of this technique, see:

G. F. ENGEN: 'A method of calibrating coaxial noise sources in terms of a waveguide standard', *IEEE Trans. Microwave Theory & Tech.*, MTT-16, pp. 636-639, Sept. 1968.

G. F. ENGEN: 'An evaluation of the 'back-to-back' method of measuring adapter efficiency', *IEEE Trans. Instr. & Meas.*, I-M19, pp. 18-22, Feb. 1970

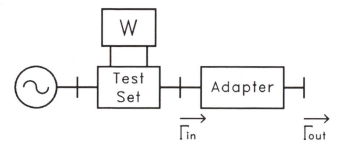

Fig. 18.3 Experimental technique for the measurement of adapter parameters

interest. From eqn. 16.5, the complex response, w, to the reflection, Γ_{in}, which obtains at the input to the adapter is given by

$$w = \frac{a\Gamma_{in} + b}{c\Gamma_{in} + 1} \qquad (18.15)$$

and from eqn. 5.5 the relationship between Γ_{in} and Γ_{out}, for the adapter may be written

$$\Gamma_{in} = \frac{\alpha\Gamma_{out} + \beta}{\gamma\Gamma_{out} + 1} \qquad (18.16)$$

where

$$\alpha = S_{12}S_{21} - S_{11}S_{22} \qquad (18.17)$$

$$\beta = S_{11} \qquad (18.18)$$

$$\gamma = -S_{22} \qquad (18.19)$$

The substitution of eqn. 18.16 in eqn. 18.15 now yields

$$w = \frac{A\Gamma_{out} + B}{C\Gamma_{out} + 1} \qquad (18.20)$$

where

$$A = \frac{a\alpha + b\gamma}{1 + \beta c} \qquad (18.21)$$

$$B = \frac{a\beta + b}{1 + \beta c} \qquad (18.22)$$

$$C = \frac{c\alpha + \gamma}{1 + \beta c} \qquad (18.23)$$

By use of a suitable calibration technique, such as that described in Chapter 16, for example, the parameters a, b, c (which characterise the reflectometer) may be determined. An application of the same (or alternative) methods at the adapter output port will yield, A, B and C. Equations 18.21 $\cdots$ 18.23 may now be solved for α, β and γ,

$$\alpha = \frac{A - bC}{a - Bc} \qquad (18.24)$$

$$\beta = \frac{B - b}{a - Bc} \qquad (18.25)$$

$$\gamma = \frac{aC - Ac}{a - Bc} \qquad (18.26)$$

Finally, in terms of the adapter scattering parameters, one has[65]

$$S_{12}S_{21} = \alpha - \beta\gamma \qquad (18.27)$$

$$S_{11} = \beta \qquad (18.28)$$

$$S_{22} = -\gamma \qquad (18.29)$$

The foregoing provides a 'complete' description or evaluation of the adapter parameters in terms of its scattering coefficients, S_{ij}. For many applications the parameter of interest is the *efficiency*, *available power ratio*, or possibly the *maximum efficiency*. Assuming the availability of sliding shorts in the different transmission lines, these may be obtained as described in Chapter 15 on *Power equation methods*.

Application of the 'TRL' calibration technique

The *through-reflect-line (TRL)* calibration technique, as described in the preceding chapter, may also be applied. Referring to Fig. 18.4, one assumes that the dual six-port has been calibrated, presumably, *but not necessarily*, by the *TRL* technique, such that Γ_1 and Γ_2 may be obtained from the power meter response. An adapter *pair* is then connected to the original test ports such that a new set of

---

[65]Ordinarily, reciprocity obtains so that $S_{12} = S_{21}$.

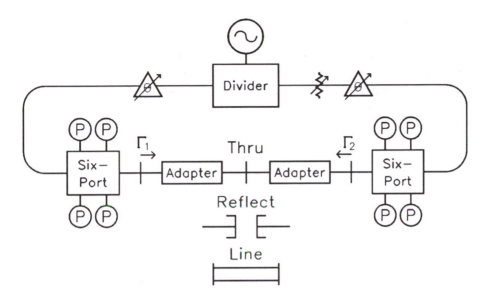

Fig. 18.4 Application of 'TRL' to adapter evaluation

'test ports' is obtained and the *TRL* procedure applied.   By comparison with Fig. 17.8, the w_1 and w_2 become Γ_1 and Γ_2, while the 'error-boxes' are the adapters to be evaluated.   As before, the complete set of scattering parameters is obtained.

Accuracy considerations

Although a detailed analysis is beyond the scope of this book, some general observations may be of value.   In many cases, as noted, the adapter *efficiency* is the parameter of greatest interest.   From eqn. 5.13, this is determined primarily by $|S_{21}|^2$ (or, assuming that reciprocity is satisfied, by $|S_{12}S_{21}|$).   Ordinarily, the magnitudes of β and γ are small in relation to that of α and similarly for b and c or B and C in relation to a and A, respectively.   From inspection of eqn. 18.24, the error in the efficiency measurement will be determined primarily by the error in the ratio A/a.   If the *TRL* calibration technique, or that suggested in Chapter 16 is employed, this error will be determined primarily by the uncertainty in the values of the 'reflection standards' (typically 'shorts' or 'opens') which are used in the calibration.   A similar observation holds for the *power equation* methods.   An inspection of eqns. 15.26, 15.28 and 15.30, indicates that the accuracy to which q_{ga}, η_{al} and η_a are obtained will be determined primarily by the accuracy to which the ratio R_1/R_2 is measured. From eqn. 9.4, R is approximately proportional to r, where r represents the magnitude of the reflection for the sliding 'short'. If nominally equal, the inevitable deviations of the sliding shorts from the assumed reflection magnitude of unity will tend to cancel. On the other hand, if there is a substantial difference in their departures from this ideal, this would represent a first order error. It is possible, however, to avoid this error by returning to eqns. 18.1 $\cdots$ 18.5 and using eqns. 18.24 $\cdots$ 18.29 to obtain an explicit value for η_{21}/η_{12}. This is done as follows:

By definition, η_{21} is given by eqn. 5.13. Assuming reciprocity, and after making use of eqns. 18.27 $\cdots$ 18.29, one has

$$\eta_{21} = \frac{|\alpha - \beta\gamma| \left[1 - |\Gamma_c|^2\right]}{|1 + \gamma\Gamma_c|^2 - |\alpha\Gamma_c + \beta|^2} \tag{18.30}$$

In order to obtain η_{12}, it is only necessary to replace β by $-\gamma$, γ by $-\beta$ and Γ_c by Γ_w. This yields

$$\eta_{12} = \frac{|\alpha - \beta\gamma| \left[1 - |\Gamma_w|^2\right]}{|1 - \beta\Gamma_w|^2 - |\alpha\Gamma_w - \gamma|^2} \tag{18.31}$$

Next, after taking the ratio of eqn. 18.30 to eqn. 18.31, one has

$$\frac{\eta_{21}}{\eta_{12}} = \frac{\left[|1 - \beta\Gamma_w|^2 - |\alpha\Gamma_w - \gamma|^2\right]\left[1 - |\Gamma_c|^2\right]}{\left[|1 + \gamma\Gamma_c|^2 - |\alpha\Gamma_c + \beta|^2\right]\left[1 - |\Gamma_w|^2\right]} \tag{18.32}$$

In particular the factor $|\alpha - \beta\gamma|$, whose magnitude is dominated by that of α, has been eliminated, and along with it the problem in making an accurate determination of $|A/a|$. It is now a simple matter to combine eqn. 18.32 with eqn. 18.5 to obtain explicit values of η_{21} and η_{12}, which may then be used in eqns. 18.1 and 18.2 to obtain an explicit value for the ratio η_c/η_w etc.[66] Although the implied procedure is a rather complicated one, it also represents a substantial improvement in potential accuracy as compared with those already described.

---

[66]By a straightforward extension of these ideas one could also devise a method for obtaining the reflection coefficient magnitude of a coaxial short in terms of that of a waveguide one etc.

Application of power equation techniques

Referring to Fig. 15.8, which has been reproduced below as Fig. 18.5, one begins by an adjustment of tuner T_X such that, as

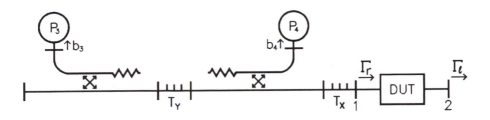

Fig. 18.5 Power equation method for measuring the ratio of adapter efficiencies

explained in Chapter 15, one has for Γ_g (which is the equivalent reflection coefficient of the measurement system) either $\Gamma_g = \Gamma_w$ or, alternatively, $\Gamma_g = \Gamma_c$. In the absence of phase response, T_Y is then used to make the system response, w, vanish when port 2 is terminated by Γ_c if $\Gamma_g = \Gamma_w$ (or, conversely, by Γ_w if $\Gamma_g = \Gamma_c$). By use of eqns. 15.51 and 15.53 and noting that in the present context $\eta_{21} = \eta_{al}$ and $\eta_{12} = q_{ga}$, one has

$$\frac{\eta_{21}}{\eta_{12}} = \frac{\left(1 - \dfrac{|w - R_{c2}|^2}{R_2{}^2}\right)}{\left(1 - \dfrac{|w - R_{c1}|^2}{R_1{}^2}\right)} \tag{18.33}$$

This result may be used in lieu of eqn. 18.32 to obtain η_c/η_w as explained above.

Chapter 19

An introduction to the description of
Noise in communication systems

The presence of noise represents the ultimate limitation in the operation of communication systems. By definition, its amplitude at any time is unpredictable (except in the statistical sense); otherwise it would be possible (in principle anyway) to 'subtract out' its effect from the signal of interest. For this reason the description of noise signals is usually by means of terms borrowed from statistics.

The purpose of this chapter is that of giving an elementary and tutorial introduction to the description of noise in communication systems and where the basic features of the problem have often been obscured by the sophisticated mathematical tools employed. The objective is to introduce and define the quantities: *mean, variance, power spectral density, stationarity* and *distribution function*, in an intuitive and elementary manner and with a minimum use of mathematics. Wherever possible, analogies will be drawn from continuous wave (CW) analysis and the basis for using the results of CW (or frequency domain) analysis to predict the operation with noise signals will be described.

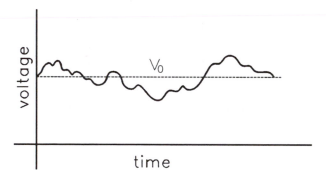

Fig. 19.1 A random noise voltage

Referring to Fig. 19.1, it will be assumed that the output from a random noise voltage generator has been recorded with the result shown. The problem is that of attempting to describe or characterise the process.

Mean and variance

Perhaps the first observation is that the fluctuations tend to occur around an average value which has been denoted V_0. By analogy one may say that the voltage has a 'DC' component. This average value, or DC component, is also called the *mean*.

Next it will be recognised that a fluctuating component is also present for which an amplitude may be assigned as is done for 'AC' signals. It will be recalled that the instantaneous value of a sinusoidal signal is continuously changing and that its average value is zero. In order to assign an 'amplitude' to a sinusoidal voltage, one usually employs the root-mean-square (RMS) approach, where the instantaneous values are squared, averaged and finally the square root taken. For a sinusoidal signal the RMS amplitude is approximately 70% of the peak value. In general the RMS definition has the property that an RMS current of a given value will deliver the same average power to a resistive load as a DC current of the same value, irrespective of the current waveform.

Returning to Fig. 19.1, the *variance* is obtained by first subtracting the *mean* (or DC) value and then finding the mean-square of the remaining waveform. In other words, the instantaneous values (after subtracting V_0) are squared and averaged. (But the square root is not taken.) If the *mean* is zero, the *variance* is thus the square of the RMS value and the square root of the variance, which is also known as the *standard deviation*, may be interpreted as the 'amplitude' of the AC component. If the voltage which has been recorded in Fig. 19.1 were impressed across a 1Ω resistor, it can be shown that the average power dissipation is the square of the *mean* plus the variance or, stated in other words, the total power is the sum of the DC and AC components, as calculated in a CW problem.

The *mean* and *variance* thus permit one to calculate the noise power, provided that the load resistance is given.

Power spectral density

The next parameter to be considered is the *power spectral density*. If, for example, one wishes to operate a radio receiver in the vicinity of an arcing motor (or noisy computer!) one may be interested in the total noise power radiated from the motor. The amount of noise power, *in the pass band of the receiver*, when it has been tuned to the desired station, however, is usually of greater interest.

The answer to this question (and others) is provided by the *power spectral density*, which is simply a measure of how the noise power output from a particular source is distributed in the frequency spectrum. Conceptually, the power spectral density may be measured by means of a very narrow band pass filter, of adjustable centre frequency and a power meter.

In Fig. 19.2, for example, it is convenient to assume a filter bandwidth of 1 Hz. The power meter reading will then be in Watts per cycle of bandwidth and, if these values are plotted for different choices of centre frequency, the *power spectral density* is obtained.

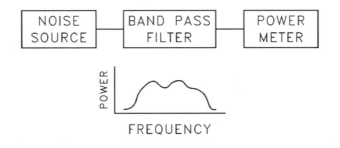

Fig. 19.2 Technique for measuring power spectral density

As a practical matter, it is usually more convenient to determine the power spectral density by indirect methods. Given a sufficient length of the record in Fig. 19.1, for example, it is possible to compute the *autocorrelation function* and then take its *Fourier transform* to obtain the power spectral density. Alternatively, it is also possible (subject to certain qualifications) to start with the *Fourier transform* of the record in Fig. 19.1 and obtain the power spectrum from the amplitude of the voltage spectrum. The former method is usually preferred, however.

Finally, it may be possible to calculate the autocorrelation function and thus the power spectral density, provided that certain details of the noise generating mechanism are known. The further development of these ideas is, however, outside the scope of this discussion.

Stationarity

As already noted, it is the essence of a noise signal that its actual values are not known in advance. It is often the case, however, that the statistical properties, i.e. *mean, variance, power spectral density* etc. are constant with time. Indeed, there would be little point in calibrating a noise source, or evaluating the noise characteristics of an amplifier, if it could not be assumed that the measured parameters were reasonably constant. A noise source for which these statistical properties is constant is called *stationary*.

Distribution function

A further method of classifying or characterising noise signals is in terms of their *distribution function*. In the noise signal shown in Fig. 19.1, the most *probable* value is V_0 and the probability that the value differs from V_0 by a given amount, x, decreases rapidly as the magnitude of x increased. In other cases, however, the picture may be quite different. If a recording were made of ignition noise, for example, one would expect to see periods of no signal interspersed with others of short duration and large amplitude. In this case (if the pulses are not too frequent) the signal level is nominally zero a large part of the time and at the peak value for most of the remainder. The amount of time the signal spends at the *mean* and other intermediate values will be very small. It is entirely possible for a signal of this type to have the same *mean, variance* and *power spectral density* as that indicated in Fig. 19.1 and yet its effect on a given communications system may be quite different. In practise, it may be desirable to take account of this in designing a circuit to detect a signal in the presence of noise.

The distribution functions commonly encountered include the *Gaussian, Poisson, Rayleigh, rectangular* and others. The Gaussian distribution is the most common and, unless otherwise specified, is usually assumed.

Application of CW analysis to noise problems

It is next desirable to say something about the basis for using the results of CW analysis to predict system operation with noise.

As is well known, a parameter of great importance in the description of communication systems is the *frequency response*. To the connoisseur of music, the current 'rock-and-roll' may have little or nothing in common with a symphony orchestra. To the audio engineer, on the other hand, the technical problems in recording or reproducing the two are substantially identical. The common feature, of course, is that both are comprised of a linear superposition of frequencies in the nominal range 20–20 000 Hz.

As previously noted, the mathematical theory upon which this conclusion is based is the *Fourier transformation*. This theory is also applicable to noise signals, but with certain qualifications which will now be described.

In the first place, one would like to begin by assuming that the noise source is *stationary* (or that its statistical properties are unchanging). At first glance, this appears innocent enough, it implies, however, that the process has been in existence throughout the infinite past and will continue into the infinite future. Obviously, no such record exists, but even if it did, it could not be represented by a frequency spectrum because the Fourier transform fails to converge and thus does not exist. (The convergence problem may be avoided by computing the *autocorrelation function* from which the *power spectrum* may be obtained and in which the phase information does not appear. This approach is thus of greater theoretical interest, notwithstanding the fact that no such noise record exists.)

As a practical matter, it is necessary to limit any attempt to obtain the properties of a noise source, from an examination of its past output, to a finite length of record. The problem is then one of attempting to determine how long the record must be in order to be truly representative. Quite obviously, if the noise source contains components with a period of a day (or year!), this will not be evident from a ten minute sample. In this case, the 'noise' is manifested as a 'drift', or 'instability' in the system parameters.

Provided that one is interested in the effect on a particular system, with certain band pass characteristics, it is usually sufficient to examine the noise source for components in this frequency range. This generally assumes, however, that the amplitudes of the components outside this range are bounded in a more or less prescribed fashion. In the final analysis, these ultimately become matters of engineering judgment. As noted in Chapter 1, 'No model is perfect $\cdots$'.

The foregoing paragraphs have considered most of the questions which emerge in attempting to apply the results of CW analysis to noise type signals. Some additional details will be given in Chapter 21. Aside from the matter of good engineering judgment, the problems are primarily of a theoretical rather than practical nature. Generally speaking, the frequency or power spectrum provides an adequate approximation for many engineering applications. Applications of these basic ideas will be made in the chapters which follow.

Chapter 20

Noise standards and radiometry

The existing techniques for evaluating the noise contributions from microwave amplifiers, mixers etc. call for the use of noise generators or sources of known parameters or characteristics. This chapter will focus on noise sources, standards and their intercomparison.

Thermal and other noise sources

Perhaps the simplest type of noise source is a resistive element (e.g. resistor, waveguide termination etc.). From the Nyquist theorem,[67] which is based on quantum thermodynamics, it is possible to associate therewith an *available* power, P_n, which is given by[68]

$$P_n = \frac{hfB}{e^{(hf/kT)} - 1} \tag{20.1}$$

and where h is Plank's constant (6.6×10^{-34}), k is Boltzman's constant (1.38×10^{-23}) and f, B and T, are the frequency, bandwidth and absolute temperature, respectively. Ordinarily, $hf \ll kT$ and eqn. 20.1 becomes,

$$P_n \approx kTB \tag{20.2}$$

This provides the basis for the use of a heated or cooled termination as a noise standard. Moreover, from a metrology standpoint, much of the prior treatment of power, its measurement, and related topics is applicable in the noise context, but with certain extensions which will be noted in due time.

Other sources of noise which are useful at microwave frequencies include the gas discharge tube, avalanche diodes and, at the lower

---

[67] H. NYQUIST: 'Thermal Agitation of Electricity in Conductors', *Phys. Rev. 32*, pp.110-113, July, 1928

[68] If the waveguide supports more than one mode, this power will be available in each of them.

203

portion of the frequency region, the temperature limited diode. Although it is possible, with varying degrees of success, to predict the noise output of these devices on the basis of the physics involved, a much higher degree of confidence is usually associated with the thermal standard. On the other hand, after their calibration against a thermal standard, these devices may play an important role as secondary or transfer standards. Ordinarily, this calibration is in terms of an 'equivalent' temperature. This is the physical temperature required in a heated (or cooled) termination which produces the same noise power output.

Two-port noise contributions

One of the major objectives of noise metrology is to evaluate the noise characteristics of two-port devices such as amplifiers. This will be the subject of the chapter to follow. This section will develop the theory associated with the use of a calibrated attenuator, as used with a noise standard (usually, but not necessarily of the thermal type), to adjust the noise temperature.

Referring to Fig. 20.1, one has a (thermal) noise source at the temperature T_g and an attenuator (usually variable) at the temperature T_a. The objective is to determine the temperature, T_s, of the

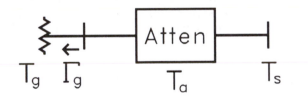

Fig. 20.1 The 'temperature' of a noise source may be reduced by the use of an attenuator.

equivalent source which obtains at the output port 2. A partial treatment of this problem will be found in Chapter 5, where the ratio of available powers (and thus noise temperatures) is given by eqn. 5.17. In the noise context, however, this result is incomplete in that *it fails to take account of the noise generated within the attenuator.* In general, the attenuation is achieved by the introduction of resistive elements and, unless at the absolute zero in temperature, these will also make a contribution to the available noise power at port 2, whose magnitude may be determined as follows.

In keeping with eqn. 15.17, the available power ratio will be denoted by q_{ga}. From eqn. 5.17 this is given by

$$q_{ga} = \frac{|S_{21}|^2 \left[1 - |\Gamma_g|^2\right]}{|1 - S_{11}\Gamma_g|^2 - |(S_{12}S_{21} - S_{11}S_{22})\Gamma_g + S_{22}|^2} \quad (20.3)$$

Although a physical impossibility, it is convenient to begin by assuming that the attenuator or two-port temperature, T_a, is zero. The temperature, T_s, is then given by

$$T_s = T_g q_{ga} \quad (20.4)$$

For other choices of T_a, it will be assumed that T_g and its contribution to T_s, remains constant. If now $T_a = T_g$, one has, $T_s = T_g$, which may also be written

$$T_s = T_g q_{ga} + T_g(1 - q_{ga}) \quad (20.5)$$

where the first term on the right is the contribution from the source, T_g, while the second is evidently that of the attenuator, whose temperature has been assumed to equal that of T_g. For other choices of T_a one has

$$T_s = T_g q_{ga} + T_a(1 - q_{ga}) \quad (20.6)$$

and where q_{ga} is given by eqn. 20.3.[69] In a practical application, T_a is frequently at the room or ambient temperature. This result provides a method of adjusting the effective temperature of a thermal noise standard, without having to adjust the physical temperature of the standard itself. In addition, in the context of the thermal standards, this result also provides the basis for evaluating certain corrections associated with the loss in the section of transmission line used to convey the noise power from the termination at the temperature T_g to that of the environment (or to the temperature T_a).

---

[69]Provided that $\Gamma_g = 0$, and that the attenuator is 'matched' (i.e. $S_{11} = S_{22} = 0$), one has $q_{ga} = |S_{21}|^2$, which is a measure of the attenuation as given by (5.10). This is the basis for a less general counterpart of eqn. 20.6 which frequently appears in the literature.

Comparison of noise sources

The calibration or intercomparison of noise source temperatures may be achieved by a determination of the total *available* noise power in a specified bandwidth and at a given frequency.[70] This calls for the use of a comparator, which is generally known as a *radiometer*.

The basic elements of a simple radiometer are shown in Fig. 20.2. As a result of the small signal (*noise!*) levels involved, an amplifier is ordinarily required. Apart from a change in amplitude,

Fig. 20.2 A simple radiometer

however, the noise signal which emerges from the amplifier has, ideally, the same characteristics as the input, which in the absence of a *DC component* has a zero mean. The amplifier is then followed by a detector, whose output is also a noise signal, *but of nonzero mean*. For the existing detector types, this mean (or DC component) is nominally proportional to the RMS value of the noise signal and thus to the square root of the input noise power or temperature. As will become apparent in what follows, however, the failure to achieve an exact proportionality in the detector does not ordinarily introduce an error.

In addition to the nonzero mean, however, there will also be a fluctuating or 'AC' component. The detector output may thus be represented as shown in Fig. 20.3. It is evident that one's ability to *estimate* the mean will be improved if one can reduce the *variance* or 'AC' component. In addition, the estimate may also be improved by the use of a longer length of record. With regard to the first of these, the variance may be reduced by increasing the *bandwidth*. The basis for this may be intuitively recognised by conceptually subdividing the existing bandwidth into increments of 1 Hz and noting that each of these contains the same *information* about the source temperature. An increase in the bandwidth is thus the counterpart of increasing the number of *samples* in an experimental attempt to estimate the mean.

---

[70]The power spectral density is assumed to be constant throughout this bandwidth.

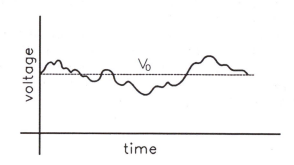

Fig. 20.3 Typical output from radiometer detector

From statistical theory, the variance of the estimated mean is inversely proportional to the number of samples and thus to the bandwidth, β. Returning to Fig. 20.2, it is also possible to decrease the variance by reducing the *cut-off* frequency of the filter that follows the detector, which in effect increases the integration time, τ, over which the response is observed.   As before, this is also analogous to an increase in the number of *samples* such that the variance is also inversely proportional to τ.

By means of the circuit in Fig. 20.1, it is possible to provide an adjustable and known temperature, T_s. Thus the temperature of an unknown noise source, T_u, may be determined as follows: The unknown source is first connected to the radiometer in Fig. 20.2 and its response observed. The unknown is then replaced by the circuit of Fig. 20.1 and the attenuator adjusted for the same radiometer response.   Ideally, the value of T_s, thus obtained, will equal the temperature of the unknown impedance.

Accuracy considerations

The above result assumes, however, that the impedances of the unknown noise source and that of 'T_s' are equal. In a more general context, the radiometer responds to a *net* noise power, whereas the desired calibration is in terms of *available* noise power or temperature. In general, these are related by 'mismatch factors' as described in Chapters 3 and 15. In addition, however, there will be a noise contribution from the amplifier, the magnitude of which also depends on the impedance of the (noise) sources. This must be explicitly recognised and accounted for if the requirements of careful metrology are to be satisfied. The theoretical basis for dealing with this problem will be developed in the chapter which follows. Another practical question pertains to the problem of recognising when the required attenuator adjustment, and thus value of T_s, has been achieved. Because one's ability to recognise this equality is obviously related to the amplitude of the AC component, the radiometer *sensitivity*, or ability to detect small changes in temperature, is usually defined to be that of the amplitude of the AC component. It can be shown that this sensitivity, ΔT, is given by[71]

$$\Delta T_{min} = \frac{(T_s + T_a)}{\sqrt{\beta\tau}} \qquad (20.7)$$

where, as will be developed in the next chapter, T_a is the noise contribution from the amplifier.

Because the radiometer is only used to recognise the *equality* of two sources, the exact detector law is unimportant, as noted above. On the other hand, any instability in the amplifier *gain* will be a source of error. Although the magnitude of this error has been greatly reduced by improved amplifier design, the problem may be

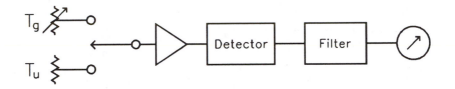

Fig. 20.4   The 'Dicke' switching radiometer

avoided by the switching or *Dicke*[72] radiometer as shown in Fig. 20.4. Here, the added feature is a fast switch which alternately samples the unknown and the standard, while the filter output is examined for a component at the switching frequency. In this way the effect of variations in the amplifier gain may be minimised.

---

[71]The integration time, τ, is not limited to that provided by the output filter. If desired, the filter output can be recorded, and further 'manual' averaging employed. As a practical matter, such a recording is generally useful as a performance monitor.

[72]R. H. DICKE: 'The measurement of thermal radiation at microwave frequencies', *Rev. Sci. Instr.*, 17, pp. 268-275, July, 1966

Chapter 21

A scattering description of

Amplifier noise

The use of noise figure or noise temperature to describe amplifier performance is a well established practise. As usually defined, these are functions of *both* the amplifier parameters *and* the source impedance. It is quite possible, however, for amplifiers to have the same noise temperature, but markedly different *sensitivities* to changes in source impedance. If optimum use is to be made of a given amplifier, its performance characteristics must be known in sufficient detail to predict its operation in an arbitrary environment, including the options which may be available in attempting to optimize the system performance and the extent or nature of the penalties which may be incurred if this is not done. This calls for a 'complete' amplifier description and the problem is one of putting this description in the most convenient form. At microwave frequencies, and in keeping with remainder of the book, this is probably in terms of the scattering parameters. It would be possible to obtain the scattering description by a reinterpretation of a number of earlier results, which are based on low-frequency circuit concepts. It will prove more instructive, however, to give its derivation from an elementary model.

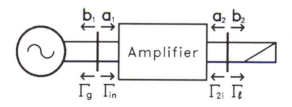

Fig. 21.1 Model for amplifier noise

The model chosen for the amplifier is a two-port which is *linear* and *active* (both in that it provides *gain* and includes *noise sources*). This device, as shown in Fig. 21.1, may be described by the equations

$$b_1 = S_{11}a_1 + S_{12}a_2 + b_{1n} \qquad (21.1)$$

$$b_2 = S_{21}a_1 + S_{22}a_2 + b_{2n} \qquad (21.2)$$

where b_1 and b_2 are the emergent wave amplitudes at ports 1 and 2, respectively, and a_1, and a_2 are the corresponding incident waves. The S_{mn} are the scattering coefficients of the two-port. Finally, the *internal* noise sources are represented by the terms b_{1n} and b_{2n}. In keeping with the arguments given in Chapter 3, this represents the most general *linear* model for the device (amplifier) of interest.

On the other hand, the use of b_{1n} and b_{2n} to represent these internal sources requires a closer examination. By hypothesis, the Fourier components included therein are confined to a 'narrow' band of frequencies which, in practical terms, means that the parameters S_{ij}, and certain others to be introduced, may be regarded as 'constant' over the bandwidth of interest. Under these conditions, but with certain limitations to be described below, the terms b_{1n} and b_{2n} may be approximately represented by sinusoidal waveforms and the complex algebra associated therewith. The 'amplitude' of b_{1n} and b_{2n} may be defined by the requirement that the power spectral densities (per cycle of bandwidth) delivered to passive and matched (reflectionless) terminations on ports 1 and 2 are given by $|b_{1n}|^2$ and $|b_{2n}|^2$, respectively. With reference to the 'phase' of b_{1n} (or b_{2n}), it is convenient to begin with a single source (b_{1n}). Let this (noise) signal be delivered to a dividing network, one portion of it delayed in time (by an amount comparable to the inverse of the frequency!) and then recombined with the other portion. To the extent that the cited approximations are valid, one may expect to obtain the same type of 'interference' as is observed with a discrete frequency or 'CW' signal and the results of CW analysis may be applied.

With the introduction of b_{2n}, however, an additional element of complexity is present. Provided that b_{1n} and b_{2n} come from 'independent' sources, and are then combined, no interference is observed. (Intuitively, this may be understood by thinking of b_{1n} and b_{2n} as individually comprised of a continuum of discrete frequencies, of infinitesimal amplitudes and random phases. As a result of the random phase stipulation, no interference is observed.) In a practical application, however, there is a strong probability that both b_{1n} and b_{2n} will include components from a common noise mechanism. Thus a measure of correlation and the potential for interference usually exists. In order to clarify this point, an analogy from discrete frequency analysis may help. As a temporary measure let

$$b_{1n} = Ae^{j\omega_1 t} + Be^{j\omega_2 t} \tag{21.3}$$

and

$$b_{2n} = Ce^{j\omega_1 t} + De^{j\omega_2 t} \tag{21.4}$$

where ω_1 and ω_2 represent two slightly different frequencies and $A \cdots D$ are complex constants. The expression for power may be written

$$P_{1n} = |b_{1n}|^2 = |A|^2 + |B|^2 \tag{21.5}$$

and

$$P_{2n} = |b_{2n}|^2 = |C|^2 + |D|^2 \tag{21.6}$$

since the time averages of the cross terms vanish. On the other hand

$$|b_{1n} + b_{2n}|^2 = |A + C|^2 + |B + D|^2 \tag{21.7}$$

Thus

$$|b_{1n} + b_{2n}|^2 = |b_{1n}|^2 + |b_{2n}|^2 + AC^* + A^*C + BD^* B^*D \tag{21.8}$$

In order to simplify the results which follow, it is possible to write

$$b_{2n} = \alpha b_{1n} + b_{2o}, \tag{21.9}$$

where

$$b_{2o} = Ee^{j\omega_1 t} + Fe^{j\omega_2 t} \tag{21.10}$$

$$\alpha = \frac{A^*C + B^*D}{|A|^2 + |B|^2} \tag{21.11}$$

$$E = \frac{B^*(BC - AD)}{|A|^2 + |B|^2} \tag{21.12}$$

and

$$F = \frac{A^*(AD - BC)}{|A|^2 + |B|^2} \tag{21.13}$$

Finally, by substitution it may be confirmed that

$$|b_{2n}|^2 = |\alpha b_{1n} + b_{2o}|^2 = |\alpha b_{1n}|^2 + |b_{2o}|^2 \tag{21.14}$$

Thus, from eqn. 21.9, it is possible, in general, to write b_{2n} as the sum of two components: The first, αb_{1n}, is completely correlated with b_{1n}, the second, b_{2o}, is uncorrelated. A similar argument may be made when b_{1n} and b_{2n} represent noise signals.[73]

From eqn. 3.10, the boundary condition imposed on b_1 and a_1 by the generator may be written

$$a_1 = b_g + b_1\Gamma_g \tag{21.15}$$

while at terminal 2

$$a_2 = b_2\Gamma_l \tag{21.16}$$

where Γ_l is the reflection of the termination which is connected to the amplifier. As indicated by eqn. 21.3, a_1, and ultimately b_2, will contain a component due to b_1 (or b_{1n}). (Conceptually the emerging noise signal, b_{1n}, is partially reflected by the generator and becomes part of the amplifier 'input' signal.) For the moment, let $b_g = a_2 = 0$, in which case $b_2 = \kappa b_{1n} + b_{2n}$, where $\kappa = S_{21}\Gamma_g$. In harmony with eqns. 21.9 and 21.14, the *functional form* for the power, P_2, delivered to the (matched) termination on port 2 is given by

$$P_2 = |b_2|^2 = |(\kappa + \alpha)b_{1n}|^2 + |b_{2o}|^2 \tag{21.17}$$

etc.

Scattering equation formulation

Returning to eqn. 21.2, it will prove convenient to let

$$b_{2n} = S_{21}a_{1n} \tag{21.18}$$

Then eqn. 21.2 may be written

$$b_2 = S_{21}(a_1 + a_{1n}) + S_{22}a_2 \tag{21.19}$$

---

[73]For a more complete discussion see H. BOSMA: 'On the theory of noisy systems', *Phillips Research Reports, Supplements*, 1967, No. 10.

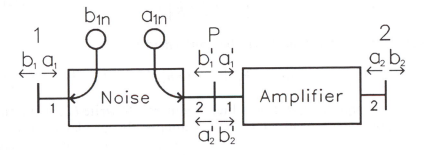

Fig. 21.2 Model for amplifier noise

It is now possible to represent eqns. 21.1 and 21.19 by means of the model given in Fig. 21.2. Here, the noise sources, a_{1n} and b_{1n}, are associated with the 'noise' two-port, and where, assuming that its scattering parameters are N_{ij}, one has $N_{11} = N_{22} = 0$ and $N_{12} = N_{21} = 1$. The 'amplifier' two-port in Fig. 21.2 is now the noise free 'equivalent' of the actual amplifier and its parameters are given by the S_{ij} in eqns. 21.1 and 21.2 or eqn. 21.19.

At the primed interface, 'P', between the noise and amplifier two-ports in Fig. 21.2 one has

$$b_2' = a_1 + a_{1n} \qquad (21.20)$$

and

$$a_2' = b_2'\Gamma_{in} \qquad (21.21)$$

where Γ_{in} is the reflection coefficient at the amplifier input. In addition, as an alternative to eqn. 21.1, one has

$$b_1 = a_2' + b_{1n} \qquad (21.22)$$

These results may now be combined with eqn. 21.15 to obtain

$$b_2' = \frac{b_g + b_{1n}\Gamma_g + a_{1n}}{1 - \Gamma_g\Gamma_{in}} \qquad (21.23)$$

In keeping with eqn. 21.9, it is possible write a_{1n} as the sum of two components,

$$a_{1n} = \xi b_{1n} + a_{1o} \qquad (21.24)$$

$$b_2' = \frac{b_g + b_{1n}(\Gamma_g + \xi) + a_{1o}}{1 - \Gamma_g \Gamma_{in}} \tag{21.25}$$

and from eqn. 21.14 one has

$$|b_2'|^2 = \frac{|b_g|^2 + |b_{1n}|^2 |\Gamma_g + \xi|^2 + |a_{1o}|^2}{|1 - \Gamma_g \Gamma_{in}|^2} \tag{21.26}$$

From eqn. 3.9, the *net* power input to the amplifier (model) at terminal P may be written

$$P_{in(net)} = |b_2'|^2 \left(1 - |\Gamma_{in}|^2\right) =$$

$$\frac{\left(1 - |\Gamma_{in}|^2\right)}{|1 - \Gamma_g \Gamma_{in}|^2} \left(P_g\left(1 - |\Gamma_g|^2\right) + |\Gamma_g + \xi|^2 |b_{1n}|^2 + |a_{1o}|^2\right) \tag{21.27}$$

where $|b_g|^2$ has been replaced by $P_g\left(1 - |\Gamma_g|^2\right)$ and P_g is the available power from the generator. At the amplifier output port (2) one has

$$P_{out} = G_{21} P_{in} \tag{21.28}$$

where G_{21} represents the amplifier gain. (An explicit expression for G_{21} is given by eqn. 5.13 but where the symbol η_{21} has been employed. This is ordinarily done to indicate that the two-port device is *passive*.)

Effective input noise temperature

By definition the *effective input noise temperature*, T_e, is that noise temperature which, if assigned to the specified source impedance, and in operation with a noise-free equivalent of the amplifier, would provide the same output noise power as the actual amplifier does when used in conjunction with a noise-free equivalent of the source impedance. In order to apply this definition, it will be assumed that the generator is a thermal noise source of temperature T_g. Thus $P_g = T_g$.

Returning to eqn. 21.27, a noise-free source results by letting $T_g = 0$ (or $P_g = 0$), a noise-free amplifier if $|a_{1o}|^2 = |b_{1n}|^2 = 0$. Application of the preceding definition thus leads to

$$T_g\left(1 - |\Gamma_g|^2\right) = |a_{1o}|^2 + |b_{1n}|^2|\Gamma_g + \xi|^2 \qquad (21.29)$$

and

$$T_e = \frac{|a_{1o}|^2 + |b_{1n}|^2|\Gamma_g + \xi|^2}{1 - |\Gamma_g|^2} \qquad (21.30)$$

It is of interest, and of substantial practical importance that $T_{e(min)}$ and the associated value of Γ_g are independent of Γ_{in} and thus the amplifier parameters and the impedance conditions which obtain at the output port (2). In keeping with its practical importance, the discussion will now focus on T_e.

The determination of the minimum value of T_e, and the corresponding value of Γ_g, may be obtained as follows: By inspection the argument of Γ_g should be chosen to equal that of $\xi + \pi$. Then, in eqn. 21.30, $|\Gamma_g + \xi|^2$ may be replaced by $\left[|\Gamma_g| - |\xi|\right]^2$ and differentiated with respect to $|\Gamma_g|$. This leads to

$$T_{e(min)} = |a_{1o}|^2 + \left[|a_{1o}|^2 + |b_{1n}|^2\left(1 - |\xi|^2\right)\right]K \qquad (21.31)$$

where

$$K = \tfrac{1}{2}\left(\sqrt{1 + F} - 1\right) \qquad (21.32)$$

and

$$F = \frac{4|a_{1o}|^2|b_{1n}|^2|\xi|^2}{\left[|a_{1o}|^2 + |b_{1n}|^2\left(1 - |\xi|^2\right)\right]^2} \qquad (21.33)$$

The realisation of $T_{e(min)}$ is obtained when $\Gamma_g = \Gamma_{opt}$ and where

$$\Gamma_{opt} = \frac{-|b_{1n}|^2\xi}{|a_{1o}|^2 + |b_{1n}|^2 + K\left[|a_{1o}|^2 + |b_{1n}|^2\left(1 - |\xi|^2\right)\right]} \qquad (21.34)$$

If Γ_{opt} is substituted for Γ_g in eqn. 21.30, T_e becomes $T_{e(min)}$. The resulting equation may then be subtracted from eqn. 21.30 to yield

$$T_e = T_{e(min)} + \frac{T_k |\Gamma_g - \Gamma_{opt}|^2}{\left[1 - |\Gamma_g|^2\right]\left[1 - |\Gamma_{opt}|^2\right]} \qquad (21.35)$$

where

$$T_k = \left[\left(|a_{1o}|^2 + |b_{1n}|^2\left[1 - |\xi|^2\right]\right)\right]\left[1 + 2K\right] \qquad (21.36)$$

This result may also be written

$$T_e = T_{e(min)} + \frac{T_k |\Gamma_{go}|^2}{1 - |\Gamma_{go}|^2} \qquad (21.37)$$

where

$$\Gamma_{go} = \frac{\Gamma_g - \Gamma_{opt}}{1 - \Gamma_g \Gamma_{opt}^*} \qquad (21.38)$$

In keeping with the treatment of 'mismatch', given in Chapter 15, Γ_{go} may be regarded as a parameter whose magnitude is a measure of the extent to which Γ_g differs from that which yields the minimum noise temperature.

Comparison with circuit formulation

In keeping with the overall purpose, it will prove useful to compare this formulation with the well known result from low-frequency circuit theory:

$$T_e = T_{e(min)} + T_o \frac{R_n}{G_g} |Y_g - Y_{opt}|^2 \qquad (21.39)$$

Here, T_o is a 'reference temperature' (typically 290° K), T_e and $T_{e(min)}$ are defined as above, Y_g is the source admittance, Y_{opt} is value of Y_g for which $T_{e(min)}$ is realised, G_g is the real part of Y_g and R_n 'characterises the rapidity with which T_e increases above $T_{e(min)}$ as Y_g departs from Y_{opt}'. Let Y_o represent a (real) reference or characteristic admittance. If one makes the substitution

$$Y_g = Y_o \frac{1 - \Gamma_g}{1 + \Gamma_g} \tag{21.40}$$

and similarly for Y_{opt}, this becomes

$$T_e = T_{e(min)} + \frac{4T_o R_n G_{opt} |\Gamma_g - \Gamma_{opt}|^2}{\left(1 - |\Gamma_g|^2\right)\left(1 - |\Gamma_{opt}|^2\right)} \tag{21.41}$$

Comparison of this result with eqn. 21.35 shows them to be equivalent (*as indeed they should be!!*) where

$$T_k = \left(|a_{1o}|^2 + |b_{1n}|^2\left(1 - |\xi|^2\right)\right)\left(1 + 2K\right) = 4T_o R_n G_{opt} \tag{21.42}$$

The 'Y-factor' measurement technique

By use of eqns. 3.16 and 21.27, eqn. 21.28 may be written

$$P_{out} = G_{21} M_{g(in)}\left(T_g + T_e\right) \tag{21.43}$$

where

$$M_{g(in)} = \frac{\left(1 - |\Gamma_g|^2\right)\left(1 - |\Gamma_{in}|^2\right)}{|1 - \Gamma_g \Gamma_{in}|^2} \tag{21.44}$$

An application of eqn. 21.43 is obtained by letting T_g alternately assume the values, $T_g = T_{h(ot)}$ and $T_g = T_{c(old)}$ while keeping Γ_g constant. If the corresponding P_{out} are denoted by P_{2h} and P_{2c}, one has

$$Y = \frac{P_{2h}}{P_{2c}} = \frac{T_h + T_e}{T_c + T_e} \tag{21.45}$$

In particular, it should be noted that the factors G_{21} and $M_{g(in)}$, which are rather complicated functions of the source and amplifier parameters, cancel in forming the ratio. From this result one has

$$T_e = \frac{T_h - YT_c}{Y - 1} \tag{21.46}$$

This is the so-called 'Y-factor' measurement technique. It provides the value of T_e for that particular value of Γ_g which is embodied in the 'hot' and 'cold' sources. Although the Y-factor technique has played a major role in the emerging art, current practise calls for a more complete description of the amplifier properties as provided, for example, by the parameters $T_{e(min)}$, T_k and Γ_{opt}. One method of achieving this is to make repeated applications of the Y-factor technique in conjunction with different values of Γ_g. Following this, it is possible to fit the data to eqn. 21.35 and thus obtain values for $T_{e(min)}$, T_k and Γ_{opt}. On the other hand, the experimental procedure tends to be a rather involved and demanding one in that it calls for four or more pairs or sets of hot and cold terminations where the impedances must be known, and be equal within the set, but differ from one set to the next. Alternatively, one could utilise a method of adjusting the source impedance provided that this does not change the effective temperature. Although the latter is no problem for a termination at ambient temperature, the variations in tuner dissipation can represent a major change to the available noise power if the cold termination is at a cryogenic temperature, for example.

As an alternative, it is possible to return to eqns. 21.27 and 21.28 and form the ratio between two values of P_{out}. Here, G_{21} cancels and the correction for changes in Γ_g may be obtained from eqn. 21.27. If this is done for a suitable collection of T_g and Γ_g, it is possible to obtain $|a_{1o}|^2$, $|b_{1n}|^2$ and ξ, from which $T_{e(min)}$, T_k and Γ_{opt} may be obtained by use of eqns. 21.31 $\cdots$ 21.36. On the other hand, $|a_{1o}|^2$, $|b_{1n}|^2$ and ξ represent an alternative characterisation of the amplifier noise characteristics. To be more specific, $|a_{1o}|^2$ and $|b_{1n}|^2$ are obviously related to the magnitude or available power associated with the internal noise sources. The challenge is that of putting this description in the most convenient form. A related problem is that of incorporating the amplifier parameters in such a way as to facilitate the analysis of a cascade connection of several noisy elements and the experimental evaluation of the noise properties of active devices. Although there are a large number of possible ways of doing this a promising one is by using the *power equation* techniques which were introduced in Chapter 15.

Power equation formulation

In order to implement the power equation formulation, it is convenient to begin with Fig. 21.3 (which was introduced as Fig. 15.7 in Chapter 15). In this model the amplifier is represented by a cascade of three two-ports whose scattering parameters are $l_{11} \cdots n_{22}$. As explained in Chapter 15, 'L' and 'N' are lossless and may be characterised, respectively, by the complex parameters d and e. (See

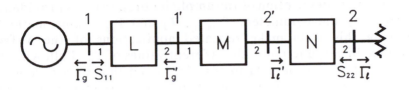

Fig. 21.3 Expanded model for amplifier noise

eqns. 15.14 $\cdots$ 15.19). On the other hand, the two-port 'M' is matched, ($m_{11} = m_{22} = 0$). From eqns. 21.1 and 21.2, the major difference between an amplifier and a passive two-port model is the addition of the source terms, b_{1n} and b_{2n}. Moreover, in order to provide gain, one has $|S_{21}| > 1$. At the same time, to avoid the potential for oscillation, it is usually necessary to stipulate that $|S_{12}S_{21}| < 1$ etc. Although the initial algebra is rather tedious, an ultimate simplification in the problem will result from making the analysis at terminal planes 1' and 2', rather than 1 and 2.

A shift in terminal planes

By a straightforward application of cascading theory, the counterpart of b_{1n} and b_{2n} at terminals 1' and 2' may be written as

$$b'_{1n} = \frac{b_{1n}\sqrt{1 - |e|^2} - m_{12}eb_{2n}\sqrt{1 - |d|^2}}{\sqrt{1 - |d|^2}\,\sqrt{1 - |e|^2}} \qquad (21.47)$$

$$b'_{2n} = \frac{b_{2n}\sqrt{1 - |d|^2} + m_{21}d^*b_{1n}\sqrt{1 - |e|^2}}{\sqrt{1 - |d|^2}\,\sqrt{1 - |e|^2}} \qquad (21.48)$$

where d and e are defined by eqns. 15.14 and 15.17.

As a counterpart to eqn. 21.7, one can now write

$$b'_{2n} = \alpha'b'_{1n} + b'_{2o} \qquad (21.49)$$

Finally, the counterparts of eqns. 21.3 and 21.4 may be written,

$$a'_1 = b'_g + b'_1\Gamma'_g \qquad (21.50)$$

and

$$a_2' = b_2' \Gamma_l' \tag{21.51}$$

where

$$b_g' = b_g \left(\frac{\sqrt{1 - |d|^2}}{1 - d\Gamma_g} \right) \tag{21.52}$$

while, from eqns. 15.34 and 15.35

$$\Gamma_l' = \frac{\Gamma_l + e}{1 + e^*\Gamma_l} \tag{21.53}$$

and

$$\Gamma_g' = \frac{\Gamma_g - d^*}{1 - d\Gamma_g} \tag{21.54}$$

It is now possible to consider the *lossless* two-ports 'L' and 'N' in Fig. 21.3 as part of the source and termination, respectively. In combination with the results obtained above, this yields the desired shift in the reference planes from 1 to 1' and from 2 to 2' and permits one to focus on the two-port 'M' which includes the noise sources. Since 'M' is *matched*, a substantial simplification in the analysis is achieved.

Development of amplifier noise model

Assuming a *matched* termination ($a_2' = 0$ at port 2'), one has

$$b_2' = m_{21}a_1' + b_{2n}' = m_{21}b_g' + \left[m_{21}\Gamma_g' + \alpha' \right] b_{1n}' + b_{2o}' \tag{21.55}$$

and letting Γ_{g2}' represent the reflection coefficient of the equivalent generator which there obtains, one has for the *available* power at port 2'

$$P_{2(avail)}' = \frac{|b_2'|^2}{1 - |\Gamma_{g2}'|^2}$$

$$= \frac{|m_{21}|^2 P_g \left(1 - |\Gamma_g'|^2 \right) + |m_{21}\Gamma_g' + \alpha'|^2 |b_{1n}'|^2 + |b_{2o}'|^2}{1 - \eta_r^2 |\Gamma_g'|^2}$$

$$\tag{21.56}$$

$$\eta_r = |m_{12}m_{21}| \qquad (21.57)$$

while P_g is the available power from the source and is given by

$$P_g = \frac{|b'_g|^2}{1 - |\Gamma'_g|^2} \qquad (21.58)$$

The value of the terminal invariant formulation is now beginning to emerge in that P_g is the same at both terminals 1 and 1', while $P_{2(avail)} = P'_{2(avail)}$. In a similar way, assuming a *passive* (noise free) termination on port 2, the available noise power from the amplifier input terminals, due to the internal noise sources, is given by

$$P_{1(avail)} = P'_{1(avail)} = \frac{|1 - m_{12}\alpha T'_l|^2 |b'_{1n}|^2 + |m_{12}\Gamma'_l|^2 |b'_{2o}|^2}{1 - \eta_r^2 |\Gamma'_l|^2}$$

$$(21.59)$$

where, as previously defined, Γ'_l is the reflection coefficient of the termination, as modified by 'N', and which obtains at terminal 2'. In many cases, both $|m_{12}|$ and $|\Gamma'_l|$ are small, thus from eqn. 21.57 η_r is small and

$$P_{1(avail)} \approx |b'_{1n}|^2 \qquad (21.60)$$

It will prove convenient to make some additional changes in notation. Let

$$T_{rev} = |b'_{1n}|^2 \qquad (21.61)$$

$$T_a = \frac{|b'_{2o}|^2}{|m_{21}|^2} \qquad (21.62)$$

and

$$\beta = \frac{-\alpha'}{m_{21}} \qquad (21.63)$$

The component of $P'_{2(avail)}$ due to P_g usually represents the signal of interest. It will be represented by P'_{2g}. Then

$$P'_{2g} = P_g \frac{|m_{21}|^2 \left[1 - |\Gamma'_g|^2\right]}{1 - \eta_r^2 |\Gamma'_g|^2} \tag{21.64}$$

By use of eqns. 21.61 $\cdots$ 21.63 the noise contribution, P'_{2n}, is given by

$$P'_{2n} = \frac{|m_{21}|^2 \left[T_a + T_{rev} |\Gamma'_g - \beta|^2\right]}{1 - \eta_r^2 |\Gamma'_g|^2} \tag{21.65}$$

Ordinarily, it is the signal (P'_{2g}) to noise (P'_{2n}) *ratio* which is of interest. This is given by

$$\frac{P'_{2g}}{P'_{2n}} = \frac{P_g \left[1 - |\Gamma'_g|^2\right]}{T_a + T_{rev} |\Gamma'_g - \beta|^2} \tag{21.66}$$

and a subject of substantial practical interest is that of choosing Γ'_g (and thus Γ_g) such that this is a maximum. This problem, however, has already been solved in a prior section, albeit in a slightly different form. In particular, an alternative definition for T_e is provided by that value of P_g for which $P'_{2g} = P'_{2n}$. From eqn. 21.66 one then has

$$T_e = \frac{T_a + T_{rev} |\Gamma'_g - \beta|^2}{1 - |\Gamma'_g|^2} \tag{21.67}$$

Thus the condition on Γ'_g for a minimum T_e is the same as for the maximum signal to noise ratio. As previously noted, $T_{e(min)}$ and the associated value of Γ'_g, are independent of Γ_l and thus the impedance conditions which obtain at the output port (2 or 2').

The counterparts of eqns. 21.31 $\cdots$ 21.34 may now be written

$$T_{e(min)} = T_a + \left[T_a + T_{rev}\left(1 - |\beta|^2\right)\right] K \tag{21.68}$$

where, as before

$$K = \frac{1}{2}\left[\sqrt{1 + F} - 1\right] \tag{21.69}$$

but where K is now given by

$$F = \frac{4T_a T_{rev} |\beta|^2}{\left[T_a + T_{rev}\left(1 - |\beta|^2\right)\right]^2} \qquad (21.70)$$

and

$$\Gamma'_{opt} = \frac{T_{rev}\beta}{T_a + T_{rev} + K\left[T_a + T_{rev}\left(1 - |\beta|^2\right)\right]} \qquad (21.71)$$

As before, if Γ'_{opt} is substituted for Γ'_g in eqn. 21.67, T_e becomes $T_{e(min)}$. After subtracting the resulting equation from eqn. 21.67 one has

$$T_e = T_{e(min)} + \frac{T_k |\Gamma'_g - \Gamma'_{opt}|^2}{\left(1 - |\Gamma'_g|^2\right)\left(1 - |\Gamma'_{opt}|^2\right)} \qquad (21.72)$$

where T_k is now

$$T_k = \left[T_a + T_{rev}\left(1 - |\beta|^2\right)\right]\left(1 + 2K\right) \qquad (21.73)$$

By use of eqn. 21.54 and its counterpart when Γ_g is replaced by Γ_{opt} etc., one can show that

$$\frac{|\Gamma'_g - \Gamma'_{opt}|}{|1 - \Gamma'_g \Gamma'^*_{opt}|} = \frac{|\Gamma_g - \Gamma_{opt}|}{|1 - \Gamma_g \Gamma^*_{opt}|} \qquad (21.74)$$

As a consequence, by use of eqn. 3.20 and in agreement with eqn. 21.35

$$T_e = T_{e(min)} + \frac{T_k |\Gamma_g - \Gamma_{opt}|^2}{\left(1 - |\Gamma_g|^2\right)\left(1 - |\Gamma_{opt}|^2\right)} \qquad (21.75)$$

and from which eqns. 21.37 and 21.38 immediately follow.

Commentary

In order to better appreciate the advantages of this alternative treatment, a further discussion of its *terminal invariant* features will prove useful. As a consequence of eqn. 21.74, and as indicated by eqn. 21.75, the coefficient of T_k in eqn. 21.73 is invariant to a shift in reference planes from 1' to 1 (or from 2 to 2') in Fig. 21.3. A similar observation may be made for the remaining parameters, T_e, $T_{e(min)}$, T_a, T_{rev}, K and $|\beta|$, which occur in eqn. 21.68, but to demonstrate this in a formal manner calls for a substantial amount of algebra.

As an alternative argument, from eqns. 21.59 and 21.61, it is possible to define T_{rev} as the noise power which obtains at the amplifier input terminals when the output has been terminated in such a way that $\Gamma'_l = 0$ and where, in order to avoid its potential contribution to T_{rev}, it is also stipulated that the physical temperature of this termination is at absolute zero. Thus T_{rev} is terminal invariant. Although the parameters, T_a and $|\beta|$, do not lend themselves as readily to a simple physical interpretation, it is possible to proceed as follows.

Let a zero temperature termination, for which $\Gamma'_g = 0$, be connected to the input port. From eqn. 21.65, the available power, P_{2o}, at port 2 is

$$P_{2o} = |m_{21}|^2 \left[T_a + T_{rev} |\beta|^2 \right] \tag{21.76}$$

Next, if a moving short (for which $|\Gamma_g| = |\Gamma'_g| = 1$) is connected to the input terminals, the maximum and minimum values for the available output powers, which are denoted P_{2M} and P_{2m}, are given by

$$P_{2M} = \frac{|m_{21}|^2 \left[T_a + T_{rev} \left(1 + |\beta| \right)^2 \right]}{1 - \eta_r^2} \tag{21.77}$$

and

$$P_{2m} = \frac{|m_{21}|^2 \left[T_a + T_{rev} \left(1 - |\beta| \right)^2 \right]}{1 - \eta_r^2} \tag{21.78}$$

The solution of eqns. 21.76 ··· 21.78 for T_a, T_{rev} and $|\beta|$ is,

$$T_a = \frac{P_{2o}}{|m_{21}|^2} \left\{ 1 + \frac{\left[P_{2M} - P_{2m} \right]^2 \left[1 - \eta_r^2 \right]}{16 P_{2o}^2 - 8 P_{2o} \left[P_{2M} + P_{2m} \right] \left[1 - \eta_r^2 \right]} \right\} \tag{21.79}$$

$$T_{rev} = \frac{\left[P_{2M} + P_{2m} \right] \left[1 - \eta_r^2 \right] - 2 P_{2o}}{2 |m_{21}|^2} \tag{21.80}$$

and

$$|\beta| = \frac{\left[P_{2M} - P_{2m}\right]\left[1 - \eta_r^2\right]}{2\left[P_{2M} + P_{2m}\right]\left[1 - \eta_r^2\right] - 4P_{2o}} \tag{21.81}$$

It is now possible to take the point of view that eqns. 21.79 $\cdots$ 21.81 provide the definitions of T_a, T_{rev} and $|\beta|$ in terms of the available powers, (which may also be experimentally observed) at the output port under the stated conditions. Because these powers, *and the experimental conditions associated therewith*, are terminal invariant, the same is evidently true for the parameters which have been thus defined.

The counterpart of eqns. 21.79 $\cdots$ 21.81 at terminal 1 is

$$T_a = \frac{P_{1o}\left[P_{1M} + P_{1m}\right]\left[1 - \eta_r^2\right] - 16P_{1o}^2 - \left[P_{1M} - P_{1m}\right]^2\left[1 - \eta_r^2\right]^2}{16P_{1o}\eta_r^2}$$

$$\tag{21.82}$$

$$T_{rev} = P_{1o} \tag{21.83}$$

and

$$|\beta| = \frac{\left[P_{1M} - P_{1m}\right]\left[1 - \eta_r^2\right]}{4P_{1o}\eta_r} \tag{21.84}$$

where P_{1o}, P_{1M} and P_{1m} may be observed at terminal 1. These results provide an alternative definition of T_a, T_{rev} and $|\beta|$. One should note, however, that eqns. 21.82 and 21.84 become indeterminate if $\eta_r \to 0$. In addition to providing these definitions, eqns. 21.79 $\cdots$ 21.84 also provide the basis for certain of the measurement techniques to be outlined below.

Since K and $T_{e(min)}$ are defined in terms of T_a, T_{rev} and $|\beta|$, (eqns. 21.68 $\cdots$ 21.70), they are also terminal invariant, as asserted above. In summary, eqn. 21.68 gives T_e as a function of Γ_g' and where the amplifier parameters are T_a, T_{rev} and β. Of these, only the argument of β fails to share the terminal invariant property. An alternative expression for T_e is given by eqn. 21.72. Here, the amplifier characterisation is effected by two real parameters, $T_{e(min)}$, T_k and the complex Γ_{opt}'. These parameter sets are interchangeable in that eqns. 21.68 $\cdots$ 21.73 may be solved for T_a, T_{rev} and β to obtain

$$T_a = \frac{T_{e(min)}\left[1 - |\Gamma'_{opt}|^2\right]\left[T_k - T_{e(min)}\right]}{T_k - T_{e(min)}\left[1 - |\Gamma'_{opt}|^2\right]} \tag{21.85}$$

$$T_{rev} = \frac{T_k - T_{e(min)}\left[1 - |\Gamma'_{opt}|^2\right]}{\left[1 - |\Gamma'_{opt}|^2\right]} \tag{21.86}$$

and

$$\beta = \frac{T_k \Gamma'_{opt}}{T_k - T_{e(min)}\left[1 - |\Gamma'_{opt}|^2\right]} \tag{21.87}$$

Alternatively, by use of eqns. 21.37 and 21.38, it is possible to express the dependence of T_e on Γ_g by the introduction of $|\Gamma_{go}|$. When this is done, Γ_{opt} takes the role of d^* in eqn. 21.54 in providing a 'reference' impedance with respect to which Γ_g is measured.

To complete this discussion, it will prove useful to compare the power and scattering equation descriptions in more detail. As already noted, both formulations lead to the same basic result which is embodied in eqns. 21.35 and 21.75. Moreover, if one compares eqn. 21.30 with eqn. 21.67, one notes that, superficially, $|a_{1o}|^2$, $|b_{1n}|^2$ and ξ have been 'replaced' by T_a, T_{rev} and $-\beta$, respectively. On the other hand, there is more involved than a mere change in notation. If one wishes to convert from one formulation to the other, the parameter d, which is a property of two-port 'L' in Fig. 21.3, is involved. To be more specific, the parameter set $|a_{1o}|^2$, $|b_{1n}|^2$ and ξ leads to $T_{e(min)}$, T_k and Γ_{opt}, via eqns. 21.31 $\cdots$ 21.34. By contrast, the use of eqns. 21.68 $\cdots$ 21.71, which call for T_a, T_{rev} and β, also provides $T_{e(min)}$ and T_k, but gives Γ'_{opt} instead of Γ_{opt}. In practical terms, this means that the reference impedance (d^*), against which the source impedance is measured, is provided by the amplifier itself rather than by some arbitrary definition of what constitutes a matched termination. This represents a potentially important practical consideration in an environment where the characteristic impedance may be poorly defined. (As an alternative, that value of 'Γ_g' for which $T_{e(min)}$ is realised could be used as a reference impedance, but measurement considerations appear to favour the use of d^*.)

Survey of measurement methods

In contrast with the preceding portions of the book, much of the material in this chapter is either 'new' or represents a recent extension of prior work. For this reason, some of the procedures to be described represent 'measurement proposals' rather than a description of established methods. Generally speaking, the merit of a given formulation (or *model!*) depends to a substantial degree on the ease (or difficulty!) in measuring the parameters on which it is based. As a rule, the measurement methods associated with the power equation or terminal invariant formulation tend to be easier to implement, although in an automated environment, this may be a moot point. On the other hand there is reason to anticipate that, as long as the ultimate measurement objective is a terminal invariant parameter, many of the measurement procedures associated with the automated network analyser, for example, may also satisfy the desired criteria, although unnecessary steps or constraints may be included in its calibration. This, however, needs to be confirmed on a case by case basis.

In order to make a more definitive comparison between the power equation and scattering formulations, and as an alternative to eqn. 21.43, the *available* (rather than *net*) power at the amplifier output port is given by

$$P_{2(avail)} = q_{ga}\left[T_g + T_e\right] \tag{21.88}$$

where q_{ga} is the available power ratio. In the scattering formulation, q_{ga} is given by eqn. 20.3, or

$$q_{ga} = \frac{|S_{21}|^2 \left[1 - |\Gamma_g|^2\right]}{|1 - S_{11}\Gamma_g|^2 - |(S_{12}S_{21} - S_{11}S_{22})\Gamma_g + S_{22}|^2} \tag{21.89}$$

By contrast, the power equation expression for q_{ga} is given by eqn. 15.32 which may be written

$$q_{ga} = \frac{|m_{21}|^2 \left[1 - |\Gamma_g'|^2\right]}{1 - \eta_r^2 |\Gamma_g'|^2} \tag{21.90}$$

As an alternative to the Y-factor technique, T_e may be obtained from eqn. 21.88 provided that T_g, $P_{2(avail)}$ and q_{ga} have been determined. Moreover, if this is done for four or more values of Γ_g (or Γ_g'), it is again possible to fit the data to either eqn. 21.35 or eqn. 21.72 and obtain $T_{e(min)}$, T_k and Γ_{opt} (or Γ_{opt}'). To the extent that the parameters in eqn. 21.90 may be more easily measured, this represents a simplification.

To continue, it is convenient to assume a source of temperature T_g at the amplifier input terminals. By combining eqns. 21.88, 21.90 and 21.67 one obtains

$$P_{2(avail)} = \frac{|m_{21}|^2}{1 - \eta_r^2 |\Gamma_g'|^2} \left(T_g\left(1 - |\Gamma_g'|^2\right) + \left[T_a + T_{rev}|\Gamma_g' - \beta|^2\right] \right)$$

(21.91)

It is possible to eliminate $|m_{21}|$ by taking the ratio of two such expressions:

$$\frac{P_{2i}\left(1 - \eta_r^2 |\Gamma_i'|^2\right)}{P_{2r}\left(1 - \eta_r^2 |\Gamma_r'|^2\right)} = \frac{T_i\left(1 - |\Gamma_i'|^2\right) + T_a + T_{rev}|\Gamma_i' - \beta|^2}{T_r\left(1 - |\Gamma_r'|^2\right) + T_a + T_{rev}|\Gamma_r' - \beta|^2}$$

(21.92)

where the subscripts 'i' (index) and 'r' (reference) have been introduced and where, for simplicity in the notation, Γ_{gr}' is just Γ_r', T_{gi} is just T_i etc.

To continue, it will be convenient to initially assume that η_r is negligibly small.[74] Then, eqn. 21.92 may be expanded to yield

$$\left[P_{2r} - P_{2i}\right]u + \left[P_{2r}|\Gamma_i'|^2 - P_{2i}|\Gamma_r'|^2\right]v - 2\left[P_{2r}\Gamma_{ix}' - P_{2i}\Gamma_{rx}'\right]x$$
$$- 2\left[P_{2r}\Gamma_{iy}' - P_{2i}\Gamma_{ry}'\right]y = T_i P_{2r}\left(1 - |\Gamma_i'|^2\right) - T_r P_{2i}\left(1 - |\Gamma_r'|^2\right)$$

(21.93)

where $u = T_a + T_{rev}|\beta|^2$, $v = T_{rev}$, $x = T_{rev}\beta_x$, $y = T_{rev}\beta_y$ and where the added subscripts 'x' and 'y' denote the real and imaginary components of Γ' and β.

Given four (or more) equations of this type ($i = 1 \cdots 4$), it is possible to solve for u, v, x, y, from which it is a simple exercise to obtain T_a, T_{rev}, β_x and β_y. Although a wide variety of choices for the Γ_i' and T_i are possible, in order to have a determinant solution, it is necessary to specify that at least one of the T_i is different from T_r (but for which $\Gamma_i' = \Gamma_r'$ is permitted) and that at least three of the Γ_i' are different from Γ_r', (for which $T_i = T_r$ is permitted).

---

[74]This will be true if $S_{12} = 0$, which is a design objective for most amplifiers.

The foregoing set of criteria thus includes the observations which are utilised in the 'Y-factor' technique for measuring $T_e(\Gamma'_r)$. Assuming that this has been done, the remainder of the task requires only the observation of the values of P_{2i} for three additional values of Γ'_i at some convenient (e.g. ambient) value of T_i. In this case, eqn. 21.93 becomes

$$P_{2r}\Big[\,|\Gamma'_i\,|^{\,2} - |\Gamma'_r\,|^{\,2}\Big]v - 2P_{2r}\Big[\Gamma'_{ix} - \Gamma'_{rx}\Big]x - 2P_{2r}\Big[\Gamma'_{iy} - \Gamma'_{ry}\Big]y$$
$$= T_i P_{2r}\Big(1 - |\Gamma'_i\,|^{\,2}\Big) - T_r P_{2i}\Big(1 - |\Gamma'_r\,|^{\,2}\Big) - T_e(P_{2r} - P_{2i})\Big(1 - |\Gamma'_r\,|^{\,2}\Big)$$

$$(21.94)$$

If one has three equations of this form, it is possible to solve for v, x and y. Then, with the help of eqn. 21.67, one can again obtain T_a, T_{rev} and β. In order to make a practical application of these results, it is necessary to measure Γ'_i, (or Γ'_g) etc. The relation, between Γ'_g and Γ_g is given by eqn. 21.54, and d may be obtained from eqn. 15.72 as explained in Chapter 15.

In theory, and referring back to eqn. 21.92, it is a simple matter to include the effect of a nonzero value for η_r. To be specific, it is only necessary to replace P_{2i} by $P_{2i}\Big[1 - \eta_r^2\,|\,\Gamma'_i\,|^{\,2}\Big]$ etc. to take account of its value. This only requires CW measurements. In practise, unfortunately, a much more difficult problem is created by the potential impact of a nonzero η_r on the instrumentation used to measure P_2. In a typical application this is usually another amplifier, which may be regarded as a 'terminating' type power meter and whose indication is proportional to *net* (rather than *available*) power. As explained in Chapter 14, these two are related by a mismatch factor, M_{gm}. For $\eta_r = 0$, M_{gm} is constant and the observed *net* power is *proportional* to the *available* power. This is the only requirement for the given procedure. When this is not true, the variation in mismatch must be taken into account. In addition, however, the terminating power meter (amplifier) will also contain its own internal noise sources which will make a contribution to its indication. This contribution, in turn, will have a dependence on Γ'_i (which now makes a contribution to its 'source impedance') as outlined above. If the amplifier being measured has sufficient gain, the noise at its output port will mask these effects.

It remains to consider the problem of measuring the (terminal invariant) amplifier parameters in an environment where impedance measurements are not available. In practise, this may be due to either the lack of suitable instrumentation, or to the nature of the physical environment (e.g. deviation from line uniformity etc.) Returning to eqns. 21.79 $\cdots$ 21.81, one has expressions for T_a, T_{rev} and $|\beta|$ in terms of P_{2M}, P_{2m}, P_{2o} and the two-port parameters, $|m_{21}|$ and η_r. Of these, the only one not amenable to evaluation, via the terminal invariant methods already described, is P_{2o} which, as already noted, is the available power at port 2 under the condition that a termination at zero temperature, and for which $\Gamma'_g = 0$, is connected to the input port. The zero temperature requirement is obviously an impossibility. However, a comparison of eqns. 21.76 and 21.67 yields

$$P_{2o} = |m_{21}|^2 T_{eo} \tag{21.95}$$

where T_{eo} is the value of T_e when $\Gamma'_g = 0$. Thus, assuming that T_{eo} can be determined (using the Y-factor method, for example) it is only necessary, in eqns. 21.79 $\cdots$ 21.81, to replace P_{2o} by $|m_{21}|^2 T_{eo}$ and one has a system of equations for measuring T_a, T_{rev} and $|\beta|$. In order to implement the Y-factor routine, however, one requires hot and cold sources (or terminations) of adjustable impedance and a technique such that the assumed $\Gamma'_g = 0$ may be both recognised and realised in each of them. A procedure for doing this was described in Chapter 15.

The noise temperature of a 'passive' two-port

To complete this discussion, it will prove useful to extend the treatment given in Chapter 20 of a two-port which, apart from internal resistive elements at the ambient temperature, T_{amb}, is passive. In conformity with the terminology introduced in this chapter, eqn. 20.6 becomes

$$T_2 = T_g q_{ga} + T_{amb}\left(1 - q_{ga}\right) \tag{21.96}$$

Returning to the definition of effective input noise temperature, the counterpart of eqns. 21.29 and 21.30 is

$$T_g q_{ga} = T_{amb}\left(1 - q_{ga}\right) \tag{21.97}$$

and

$$T_e = T_{amb}\frac{1 - q_{ga}}{q_{ga}} = T_{amb}\left(\frac{1}{q_{ga}} - 1\right) \tag{21.98}$$

The counterpart of eqn. 21.76 may be obtained from eqn. 21.96 with $T_g = 0$ and $q = q_{max}$,

$$P_{2o} = T_{amb}\left[1 - q_{max}\right] \qquad (21.99)$$

From eqn. 15.32 the value of q_{ga} for a moving short is zero and by further use of eqn. 21.96,

$$P_{2M} = P_{2m} = T_{amb} \qquad (21.100)$$

The substitution of these results in eqns. 21.79 $\cdots$ 21.81 yields

$$T_a = T_{amb}\left(\frac{1}{q_{max}} - 1\right) \qquad (21.101)$$

$$T_{rev} = T_{amb}\left(1 - \frac{\eta_r^2}{q_{max}}\right) \qquad (21.102)$$

and

$$|\beta| = 0 \qquad (21.103)$$

Equation 21.102 may also be written

$$T_{rev} = T_{amb}\left[1 - q'_{max}\right] \qquad (21.104)$$

where the prime indicates a propagation in the reversed direction. Moreover, if reciprocity obtains, which for a passive two-port is often the case, $q'_{max} = q_{max}$. From eqns. 21.68 and 21.71, one then has

$$T_{e(min)} = T_a = T_{amb}\left(\frac{1}{q_{max}} - 1\right) \qquad (21.105)$$

$$\Gamma'_{opt} = 0 \qquad (21.106)$$

and eqn. 21.72 becomes

$$T_e = \frac{T_{amb}}{q_{max}}\left\{\left[1 - q_{max}\right] + \frac{\left[1 - \eta_r^2\right]|\Gamma'_g|^2}{1 - |\Gamma'_g|^2}\right\} \qquad (21.107)$$

Appendix

The simplification which is provided by complex variable notation in circuit theory is well known and the practise is widespread if not universal. However, if, for example, one attempts to determine the (*complex*) impedance conditions which must obtain for maximum power transfer, a problem is encountered in that the expression for power includes the absolute values of certain functions of the impedances involved and is thus 'nondifferentiable'. Although it is always possible to eliminate this difficulty by reverting to a *real* notation, this may also represent a substantial increase in the effort required to obtain the derivatives. As an alternative, it is possible to replace $|z|^2$, for example, by the product of z and z^*. These, in turn, may be written as functions of two independent *real* variables x and y.

Given an analytic function F of the complex variables z, z^*, such that $F(z, z^*)$ is real for all choices of z and z^*, it will be shown that a necessary and sufficient condition for an extremum of F is that

$$\frac{\partial F}{\partial z} = 0 \qquad (A.1)$$

and in which z^* is assumed to be constant.

It will prove convenient to introduce some additional notation:

$$z = u = x + jy \qquad (A.2)$$

$$z^* = v = x - jy \qquad (A.3)$$

where x and y are the real and imaginary components of z.

Then

$$F = f(u, v) \qquad (A.4)$$

and, following the rules for differentiation:

232

$$dF = \frac{\partial f}{\partial u}\,du + \frac{\partial f}{\partial v}\,dv \tag{A.5}$$

$$dF = \frac{\partial f}{\partial u}\,(dx + j\,dy) + \frac{\partial f}{\partial v}\,(dx - j\,dy) \tag{A.6}$$

$$= \left[\frac{\partial f}{\partial u} + \frac{\partial f}{\partial v}\right]dx + j\left[\frac{\partial f}{\partial u} - \frac{\partial f}{\partial v}\right]dy \tag{A.7}$$

To simplify the notation, let

$$\frac{\partial f}{\partial u} = \alpha, \text{ and } \frac{\partial f}{\partial v} = \beta$$

Then eqn. A.7 becomes

$$dF = (\alpha + \beta)dx + j(\alpha - \beta)dy \tag{A.8}$$

By hypothesis, F is *real* for all choices of x and y. Therefore dF must be *real* for arbitrary dx and dy. This is only possible if $(\alpha + \beta)$ is a *real* number and $(\alpha - \beta)$ is *imaginary*. This leads to the following equations:

$$\alpha + \beta = \alpha^* + \beta^* \tag{A.9}$$

$$\alpha - \beta = -\alpha^* + \beta^* \tag{A.10}$$

from which

$$\alpha = \beta^* \tag{A.11}$$

From eqn. A.11, if α vanishes, the same must be true of β and eqn. A.8 shows that this is sufficient to ensure $dF = 0$. Conversely, it is easily recognised from eqn. A.8 that if dF vanishes for arbitrary dx and dy, this requires $\alpha = \beta = 0$.

This completes the proof.

Index